T0091702

Springer Series on Environmental Management

Series Editors
Bruce N. Anderson
Planreal Australasia, Keilor, Victoria, Australia

Robert W. Howarth
Cornell University, Ithaca, NY, USA

Lawrence R. Walker
University of Nevada, Las Vegas, NV, USA

For other titles published in this series, go to
www.springer.com/series/412

Douglas J. Spieles

Protected Land

Disturbance, Stress, and American
Ecosystem Management

 Springer

Douglas J. Spieles
Associate Professor of Environmental Studies
Denison University
Granville, OH 43023
USA
spielesd@denison.edu

ISSN 0172-6161
ISBN 978-1-4419-6812-8   e-ISBN 978-1-4419-6813-5
DOI 10.1007/978-1-4419-6813-5
Springer New York Dordrecht Heidelberg London

Library of Congress Control Number: 2010933785

© Springer Science+Business Media, LLC 2010
All rights reserved. This work may not be translated or copied in whole or in part without the written permission of the publisher (Springer Science+Business Media, LLC, 233 Spring Street, New York, NY 10013, USA), except for brief excerpts in connection with reviews or scholarly analysis. Use in connection with any form of information storage and retrieval, electronic adaptation, computer software, or by similar or dissimilar methodology now known or hereafter developed is forbidden.
The use in this publication of trade names, trademarks, service marks, and similar terms, even if they are not identified as such, is not to be taken as an expression of opinion as to whether or not they are subject to proprietary rights.

Cover photograph: J.C. Hidalgo/Fotolia.com.

Printed on acid-free paper

Springer is part of Springer Science+Business Media (www.springer.com)

Preface

By many measures, Earth's ecosystems are stressed. Actually, it may be more accurate to say that Earth's *remaining* ecosystems are stressed. The fact is that most of the planet's biomes support only a fraction of the biological communities they once did, primarily because humans have converted large areas of land to alternate uses. More than two-thirds of the global temperate forests, half of the grasslands, even a third of desert ecosystems have been conscripted for human uses like agriculture, construction, harvest and extraction. Cultivation alone covers a quarter of the habitable terrestrial surface. Aquatic ecosystems have not fared any better. An estimated half of the world's wetlands are gone, particularly those of coastal regions or on arable land. About a fifth of the coral reefs and a third of the mangrove swamps of a century ago have been lost in just the last few decades. The volume of water impounded by dams quadrupled over the same period – it now far exceeds the volume of water in unimpeded rivers (Millennium Ecosystem Assessment 2005; Mitsch and Gosselink 2007).

So any assessment of ecosystem status is necessarily an analysis of fragments and remnants, and many of these are degraded by one or more anthropogenic stressors. Agriculture and development have resulted in erosion and soil impoverishment; fertilizer use and waste disposal have lead to eutrophication of aquatic ecosystems; irrigation and overgrazing have rendered land barren. The list goes on. These stressors coupled with overharvest and habitat loss have contributed to an estimated 1,000-fold increase in Earth's baseline extinction rate (Millennium Ecosystem Assessment 2005). Invasive species have responded explosively to this displacement, and their introduction and expansion have altered the rules of competition. Truly, the ecological picture is bleak. This is all the more alarming because ecological systems provide many services on which humans depend. Food, fiber, and fuel, access to clean water and air, regulation of environmental processes, and even our sense of cultural legacy and wellness are dependent directly or indirectly on ecosystems.

In the relatively brief existence of the United States, ecosystem conversion and degradation have been acute. The ecological crisis may have reached a boiling point in the 1960s, but dismay over the detrimental effects of expansion and industrialization on the nation's ecosystems was evident long before. Marjory Stoneman Douglas brought the splendor and crisis of the Everglades to the world's attention in the 1940s, about the same time that Aldo Leopold taught us to think like a mountain.

Before them, John Muir did for the ecosystems of the American west what George Perkins Marsh and Henry David Thoreau had done for the nature preservation movement in the mid-nineteenth century. And yet, the dominant attitude for much of America's history was one of contempt for wilderness, accompanied by rampant development and exploitation that was at best tempered with a Gifford Pinchot-inspired conservation ethic. But attitudes shifted rather abruptly in the mid-twentieth century, precipitating a mandate for ecosystem protection. Of course, ecosystem protection was not a new idea. The first American forest protection measures were established in 1626; the first community forest reserve in 1710; the first national timberland in 1799; the first national park in 1872; the first state park and first state forest in 1885; and the first land trust in 1891 (Jensen and Guthrie 2005). But since the environmental revolution of the 1960s, federally protected acreage has increased tenfold, state-protected parks, natural areas, and forests have increased by a third, and land protected by private organizations has grown by an astonishing factor of 60 (Brewer 2003; Jensen and Guthrie 2005; Vale 2005). This land-protection renaissance has occurred largely in the spirit of preserving the ecosystems we still have and restoring those that we have lost.

Over the last century or so – while the nation has been trying to decide whether it loathes or loves its ecosystems – ecologists have been debating the mechanisms by which ecological systems assemble and function. At the heart of the debate is the question of whether ecosystems exist and develop as discrete, holistic units or whether they are simply coincidental and temporary associations of individual species. On one level this is purely an academic issue. But it also has important practical ramifications, particularly given the state of ecological degradation in which we find ourselves. It is a question that is relevant to our recent enthusiasm for ecosystem protection. Here is the dilemma: If ecosystems occur naturally as stable units with characteristic structure and function, then our efforts to protect critical ecological services must be aimed at the preservation of ecosystems in their natural state. On the other hand, if ecosystems have no stable state or characteristic composition, if instead they are ephemeral in space and time, then our ecosystem management must give priority to shifting populations and variable function.

Prevailing opinion on the nature of the ecosystem has evolved over the years in a way that is reminiscent of our vacillating national opinion on the value of wilderness. Due in part to early European influence, conservationists and preservationists in the United States have long been partial to the holistic view of the ecosystem as a unit. Various analogies have been used to describe this mindset: the ecosystem as a superorganism; the climax state; the self-regulating machine; the homeostatic entity; the self-maintained domain of attraction. In general, they all portray the ecosystem as a biological community with the ability to persist in a stable state by virtue of regulatory internal feedback mechanisms. Ecological disturbances, like flood, fire, or storm might disrupt the stable state, but in the holistic view a healthy ecosystem is *resilient*, meaning that it will return to a stable, optimal equilibrium if given the chance. An unhealthy ecosystem – one without characteristic species in appropriate abundances, for example – may gravitate toward an undesirable alternate state. To preserve ecosystem services, the holist suggests, such an alternate state is something that the ecosystem manager must guard against.

The individualistic view, historically the minority opinion in the American conservation movement, has recently gained evidence and support as a non-equilibrium concept of the ecosystem. Species, according to this view, respond individually to fluctuating environmental conditions. They occur in a shifting mosaic of successional patches that have little to do with the boundaries or labels we place upon the greater ecosystem. Stability, balance, the climax community, the domain of attraction – all are human perceptions of pattern in the noise of nature. The idea of ecosystem preservation loses some meaning in nonequilibrium ecology, for the structure, composition and function of each system are by their nature subject to change. Indeed, it casts doubt on our national effort to preserve ecosystems in form and function. How are we to maintain threatened species, vital ecological services, and our ecological legacy in a coherent state if ecosystems are not coherent? In the words of leading holist E.P. Odum, if you believe that nature is a continually shifting quilt of patches, "then there's no order, and why bother about conservation?" (Chaffin 1998).

This book is about ecological protection and management in the face of our changing concept of the ecosystem. In the first two chapters I place current examples of ecosystem protection in juxtaposition with historical ecosystem concepts, particularly the holistic and individualistic views. After this background, the first half of the book is devoted to the holistic and individualistic ways in which we conceptualize the ecosystem – including ecological integrity, health, stability, and resilience amidst disturbance, stress, and invasion. I then turn to ecosystem management in practice. In particular, I use examples of microbial, forest, grassland, freshwater, and saltwater ecosystems to evaluate the application of theory. My purpose is to clarify the disparate academic views on the ecosystem and to reconcile those views with applied ecosystem management.

If our long history of ecological destruction and degradation can teach us anything, it is that we are dependent upon the individual and collective function of other species on this planet. We now understand many things about the ways in which species associate and respond to stress and disturbance. Given our reliance on Earth's greater biological community, it would behoove us to apply our best understanding to the ways in which we protect these things we call ecosystems.

Granville, OH   Douglas J. Spieles

References

Brewer, R. 2003. Conservancy: The Land Trust Movement in America. Lebanon: University Press of New England.
Chaffin, T. 1998. Whole-Earth Mentor. Natural History 107:8–10.
Jensen, C., and Guthrie, S. 2005. Outdoor Recreation in America. Champaign: Human Kinetics Publishers.
Millennium Ecosystem Assessment, Ecosystems and Human Well-being: Synthesis. 2005. Washington: Island Press.
Mitsch, W. J., and Gosselink, J. 2007. Wetlands. Hoboken: John Wiley & Sons.
Vale, T. R. 2005. The American Wilderness: Reflections on Nature Protection in the United States. Charlottesville: University of Virginia Press.

Acknowledgments

This work was made possible by a Robert C. Good Faculty Fellowship awarded by Denison University in 2009. I am honored to have been granted this fellowship and grateful for the opportunity it has provided. I thank the many colleagues, friends and family who have offered advice, encouragement and support, and I particularly appreciate the constructive criticism of a number of anonymous reviewers throughout the course of the project. I thank the editors of the Springer Series on Environmental Management for their thoughtful consideration and general assistance. I thank the Denison University Librarians for assisting me with anything and everything. And, most importantly, I thank Beth, Jackie and Adam for being a constant source of inspiration, encouragement, and love.

Contents

Chapter 1
Four Ecosystems, Four Questions

Oak Openings, Ohio

In the sandy savannas around the Great Lakes there lives a tiny butterfly called the Karner blue. It is a delicate beauty, but to see it you have to know where and when to look. The life of an adult is short – less than 2 weeks – and there aren't as many as there used to be. In the past few decades the Karner blue population has dropped by 99%, and they are now found only in tiny remnants of their former range (Grundel et al. 1998). In part, the precipitous decline of this species is related to its feeding habits. The larvae of the Karner blue feed only on one plant, the wild lupine, and only in the northern portion of the wild lupine's range. The butterfly is also preferential to grasslands with partial tree canopy, and this sensitivity to mixed sun and shade further limits its available habitat.

Just southwest of Lake Erie is an area called the oak openings where such habitat was once relatively common. The unique characteristics of the oak openings begin with the soil; the region is underlain partially by impervious clay and partially by porous sand. The result is ideal for the Karner blue: a flood-prone woodland fringed by comparatively dry, sparse grassland. But suburban sprawl, agriculture, and other land uses have reduced the oak openings to small remnants scattered around the northeastern and upper midwestern United States (Brewer and Vankat 2006). And even the remnants aren't pristine. Fragmentation and pollution are ecologically stressful, and disturbances that maintain the savanna, particularly fire and grazing, have historically been suppressed by humans. Without these periodic disruptions, successional woody species out-compete the wild lupine and threaten the Karner blue. In 1988, this small butterfly of Ohio's oak openings was driven to local extinction (Tolson et al. 1999).

Ohio's oak openings originated with the glaciers that departed from the Great Lakes region about 14,000 years ago. Long before these sand dunes fringed the hardwood forest, the region was a great lake itself – really an extension of present day Lake Erie, but much larger. By one estimate it was 230 ft deeper than Lake Erie is today (Goldthwait 1959). The glacial meltwater that fed the lake was prevented from draining eastward to the Atlantic by massive ice dams, and so the lake grew in volume until it found an outlet in the Mississippi River basin to the west.

D. J. Spieles, *Protected Land*, Springer Series on Environmental Management,
DOI 10.1007/978-1-4419-6813-5_1, © Springer Science+Business Media, LLC 2010

At its peak the glacial lake covered much of today's northwest Ohio, southeast Michigan, and northeast Indiana. The constant stream of meltwater from the retreating glaciers eroded the surrounding landscape, and millions of tons of sediments entered the glacial lake and eventually settled on the lakebed, which would one day be the basis for a flat, impermeable landscape fringed by sandy beaches.

The glacial lake was no fleeting feature of the landscape – it existed for nearly a thousand years and became a thriving ecosystem in its own right. At first, it was a barren pool of cold, muddy water, but living things were quickly claiming environments that the glaciers surrendered. The first colonizers to arrive in the lake likely included microorganisms and seeds that were blown by the wind or transported by birds, landing by chance in the cold, turbid water. Surrounding the lake was glacial tundra, with shallow pools of meltwater and sparse vegetation stunted by permafrost and glacial winds. Over centuries of harsh conditions the diversity of living things in and around the lake gradually increased, and soon the lake teemed with fish, while mammoth, musk oxen, and caribou browsed along its shore. Through both life and death these organisms added a rain of organic detritus to the settling silt, contributing to the thick muck that would one day make this an imposing landscape for human settlers (Teller 1987).

After a thousand years of colonization and development, the glacial lake suddenly and catastrophically disappeared. The retreating glacier exposed an eastern outlet that was a great deal lower in elevation than the glacial lake. The lake lost 90% of its volume in little more than a century, reduced to a small puddle in the footprint of today's Lake Erie (Teller 1987). All that remained of the glacial lake were the remnant beaches and the basin floor. As the lake bed dried and warmed there was a rush for colonization by terrestrial plants and animals. The odds-on favorites for invasion were plants in the poll position: those that had been established along the former shoreline, which could most easily distribute seed into the drained lake basin. Sedges and grasses quickly invaded, but the real winners of this ecological lottery were cone-bearing trees that had been migrating northward, stalking the glacier in its retreat. Thus the former lake rapidly became a coniferous swamp. But this had happened before. In fact, the forest and glacier had been playing this game of cat-and-mouse for a geologic age. Prior to this most recent glaciation, and perhaps during many interglacial periods, this region had been a coniferous forest; after each establishment, advancing glaciers once again plowed through the trees and devastated the ecosystem. Ancient spruce logs have been found in present-day layers of glacial till, evidence of forests that matured and perished as dictated by the glaciers (Goldthwait 1959). And now, for a time, the spruce forest had returned.

But the ice was not finished. After five centuries of forest growth, the glaciers re-advanced and cut off the drainage routes to the east. Lake levels rose again as quickly as they had once fallen, and the forest was drowned. The region was once again a lake, this time for about 600 years. It was an ecologically chaotic time. On the southern shore, with dogged persistence, was the coniferous forest, while glacial tundra existed to the north. The tundra communities of mammoth, caribou and bison were thus in close proximity with the mastodon, stag moose, and giant sloth of the forest.

Wolves played no favorites and ranged throughout both environments. To make matters even more intriguing, the lake that now existed had been open, in alternating fashion, with both the Atlantic and Mississippi outlets. This allowed for aquatic migration from both directions, making the glacial lake an ecological melting pot in which species from the eastern seaboard co-existed alongside fish from the Mississippi River basin. Eventually, as the ice finally retreated northward and Lake Erie reached its modern level, the trees reclaimed the mud (Pielou 1991).

The ecosystem that would become known as the oak openings thus came about in what can only be described as a series of ecological convulsions – ice advance and retreat, colonization, extinction, and re-colonization; flood, permafrost, and rivers reversing flow direction. The ultimate factors governing the development of the oak openings include the glacial ice and the rock and sediment that it eroded and carried, the climate and wind patterns that deposited organisms, the regional topography, and the periodic drainage and flooding. Living things also played a key role in this ecological development – from microorganisms to megaherbivores, living things altered the soil and water conditions and contributed their biomass to the ecosystem. But there was no single ecosystem. In its first 2,000 years since glacial retreat the region had in fact been many unique ecosystems, each with distinct limitations, opportunities, and residents.

Even after the convulsions of glacial advance and retreat had finally ended, there was still tremendous ecological change. As the climate warmed over the centuries, plant and animal species migrated northward. The spruce swamp became a pine swamp, and then a willow-poplar or elm-ash-maple or oak-hickory forest, depending on location (Sampson 1930). On the sandy beaches of the glacial lakes, which were slightly higher and drier than the old lake bed, oak woodland communities assembled. Dry years and indigenous humans encouraged fires on the ridges, effectively converting the sandy uplands to prairies and oak savannas. Animal communities were also transient over the centuries. The mastodon, the stag moose, and the giant sloth all migrated away from or were hunted out of the area and eventually became extinct, replaced by black bear, deer, elk and bison. These, too were extirpated by the late nineteenth century amid massive human clearing, drainage and development that would eventually consume nearly every trace of the glacial lake and its sandy ridges (Mayfield 1962).

No one knows exactly when the Karner blue came to reside in Ohio's oak openings. But we do know that ecological change and anthropogenic stress eventually rendered its habitat unsuitable. In 1992, a partnership of conservation organizations began working to reintroduce the Karner blue to Ohio (Tolson et al. 1999). The focal point for this restoration was a small patch of oak openings that looked as they might have prior to European contact: the 750 acre Kitty Todd Preserve. Actually, the preserve might be thought of as a collection of habitats. There are grasslands here, including wet prairies that would soak your boots if you walked through them in the spring. In other places there are oak savannas that transition into forests and then, with only a few centimeters change in elevation, to treeless sand dunes. Just off the dunes you'll find patches of swamp forest, shadows of those that once made the region almost impassible and uninhabitable for humans. So there is diversity of habitat in this small preserve, which enables it to support dozens of rare plants and animal species. Prior to 1988, this included the Karner blue.

The effort to bring it back has required a great deal of ecological intervention – managers have used prescribed burning, mowing, herbicides, and manual labor to remove nonnative species and to prevent the encroachment of woody species into the prairie-savanna. They have reintroduced native plants to re-establish an appropriate prairie community and oak canopy, and most critically have propagated the wild lupine. Captive-reared Karner blues were first released in 1998, and in the last decade the species has made a modest comeback. Thus with a great deal of effort the oak openings have been reconstructed and preserved in northwest Ohio. With continued maintenance – to keep the openings from becoming invaded by woody species – the habitat of this tiny butterfly may endure.

The restoration of a wild-breeding Karner blue population in the remnant oak openings of Ohio is a remarkable achievement. The unique habitats and rare species of Kitty Todd Preserve are truly ecological treasures. And yet, in historical perspective, it is clear that these habitats, these species, are only the current permutation of endless change. Along with this legacy of change we inherit the responsibility of protecting and preserving these spectacular ecosystems. But can an ecosystem be preserved? What can it mean to preserve something that is in a constant state of change?

Kissimmee River, Florida

Far to the south of Ohio's oak openings, an ecological system exists in a state that is far from preserved. The Kissimmee River in central Florida is infamous for its story of ecosystem management gone wrong. The Kissimmee once meandered over 100 miles through a flat, wide floodplain on its journey from Lake Kissimmee to Lake Okeechobee. It was a slow, sluggish river, but during wet seasons it delivered more water to Lake Okeechobee than the Lake could discharge. Consequently, the backflow forced the Kissimmee River out if its banks and into the floodplain (Warne et al. 2000). These flood events and the surrounding topography made the Kissimmee a unique river ecosystem, primarily because of the scale and duration of its flooding. Historically, Kissimmee floodwaters filled some 35,000 acres of marshland, which then slowly released water back into the river during drier times of the year. The floodplain wetlands were an ecologically important part of the river; they were zones of nutrient and sediment exchange and areas of incredible biological diversity. Unfortunately, the flood regime was incompatible with human development in the region. By the late nineteenth century, a network of drainage ditches was removing floodwater from the land's surface, and by 1925 Lake Okeechobee was surrounded by flood control structures (Koebel 1995).

Regional human population grew in the twentieth century, and with catastrophic, hurricane-induced flooding there was increasing pressure to control the Kissimmee. In response, the US Army Corps of Engineers and the South Florida Flood Control

District (now the South Florida Water Management District) embarked on a decade-long channelization project. By 1972, the Kissimmee was converted into a 75 m wide, 9 m deep, 90 km drainage canal called C-38 – a channel much straighter, wider, deeper, and more uniform than the former river. The canal was equipped with six water control structures that effectively converted the flowing river into five reservoirs of stagnant water (Whalen et al. 2002). The floodplain wetlands were drained, creating area for agriculture and development that were reasonably well protected from flooding.

Despite these "improvements," the environmental and ecological consequences of channelization and drainage became apparent almost immediately. The Canal was designed to move excess water out of the region as quickly as possible, which it did quite effectively. As a result, the flow became flashy, with brief periods of high flow and long periods of no flow. The slow, continuous release of water that formerly came from the floodplain wetlands had been eliminated. The flow had also been altered seasonally – the greatest flow in C-38 was in June and July, historically the lowest period of flow in the unchannelized Kissimmee. The water in C-38 was also of lower quality. Excessive amounts of nutrients from floodplain agriculture and development were being transported directly into Lake Okeechobee, which quickly became thick with algae. As the algal mass decomposed, dissolved oxygen levels in the lake plummeted, along with the lake's biota. Of course, the Kissimmee flood-plain wetlands were no longer receiving floodwaters, and many had been converted into alternate land uses. The former marshland was reduced to a fraction of its former area, accompanied by declines in wildlife breeding, feeding, nesting, and growth. Groundwater, formerly recharged by the water flowing slowly over the floodplain, began to decrease in quantity and quality (Whalen et al. 2002).

Twenty years was enough to confirm that the Kissimmee River channelization project had been an unmitigated ecological disaster. In 1992, the United States Congress enacted the Water Resources Development Act, which provided for the restoration of the Kissimmee. The project has resulted in the backfilling of 22 miles of C-38 and the re-meandering of a portion of the river through its former flood-plain. Much of this land had become privately owned, so the state of Florida has acquired more than 100,000 acres in the region of Lake Kissimmee and the Kissimmee River valley, where more than 26,000 acres of wetland are to be restored. In addition, the restoration plan involves the removal of two of the six water control structures in C-38, and the flow in the river is to be returned to historic characteristics. The restoration is expected to be completed in 2011, with 5 additional years of monitoring to "ensure restoration success" (De Luise 2006).

And what is restoration success? Those involved with the restoration of the Kissimmee have given this question a great deal of thought. The South Florida Water Management District has developed a list of 25 "Restoration Expectations" that will be used to gauge the success of the Kissimmee River project (Anderson et al. 2005). Generally, the expectations fall into four categories: hydrology, water quality, habitat structure, and biological communities. The expectations are based on the best data available from pre-channelization conditions; these are the refer-ence conditions for the restoration effort. In short, the restoration may be deemed successful if these conditions are met.

Hydrologically, the expectation is that the restored ecosystem will be a continuously flowing river that varies according to seasonal climatic variation. Some aspects of the hydrologic expectations are incredibly specific – for example, the "river channel stage will exceed the average ground elevation for 180 days per water-year and stages will fluctuate by 3.75 ft" and "mean velocities within the main river channel will range from 0.8 to 1.8 ft/s a minimum of 85% of the year." The corresponding river morphology, too, has specific expectations; for example: "point bars will form on the inside bends of river channel meanders with an arc angle >70°" (Anderson et al. 2005). The water quality expectations and habitat characteristics have likewise been engineered. There are specifications for the type and width of vegetation beds within the channel and the percent cover of specific wetlands plant communities within the floodplain. One gets the impression that the restoration "endpoint" is expected to be achieved with the precision of a machine.

These "structural" characteristics of flow regime, topography, and morphology may lend themselves to precise restoration specifications, but what about the living organisms? Though the restoration has not focused on individual species, there are certainly parameters of expectation. What types of invertebrates will be the most common in the flowing water of the river? What will the species richness and diversity be in the broadleaf marsh community? How many species of reptile, amphibian, fish and birds will occur in the restored ecosystem? All are specified. To wit: "mean annual density of small fishes (fishes <10 cm total length) within restored marsh habitats will be >18 fish/m^2…mean annual dry season density of long-legged wading birds (excluding cattle egrets) on the restored floodplain will be >30.6 birds/km^2… and winter densities of waterfowl within the restored area of floodplain will be 3.9 ducks/km^2" (Anderson et al. 2005).

To the authors' credit, these are probably the best researched set of river restoration goals in the history of river restoration. And it is admirable to expect that the restored Kissimmee River will match the pre-channelization river so completely. But this is a living system, after all. Is it realistic to expect it to fall neatly into such a rigid set of criteria for success? Is that really the point of restoration?

Tallgrass Prairie, Kansas

One hundred and forty million acres of tallgrass prairie once occupied the eastern extent of the Great Plains. Maintained by climate, fire, and grazing, it both formed and responded to the Native American culture that called it home. Some 50 million bison grazed along with elk, pronghorn, and deer as they roamed the open expanse. But then came the rancher, the plow, and the rifle, and the prairie began to disappear with astonishing speed. By the mid-twentieth century it was nearly gone, and today only about 4% of the pre-European American settlement tallgrass prairie remains. A good portion of the remnants exist in two Kansas preserves: the Tallgrass Prairie National Preserve and Konza Prairie Preserve (Savage et al. 2004).

Tallgrass Prairie National Preserve was created in 1996 in the Flint Hills region of eastern Kansas. The Preserve consists of nearly 11,000 acres that are managed by the National Park Service but owned by the Nature Conservancy. The rolling hills include some of the last and most pristine unplowed grassland in the country. Inhabited in at least a transient way by humans for at least 10,000 years, the Kansas grasslands were a site of nomadic hunting, plant domestication and horticulture (Jones 1999). As European American settlers encroached, much of the tallgrass prairie was converted to cultivation and grazing. The Flint Hills region has thin, rocky soil that was deemed unsuitable for cultivation and thus spared the plow – though it has been the site of intensive cattle grazing since about 1880 and will be at least through 2030, when the current lease expires.

Grazing and fire are two of the most important factors that shaped the pre-settlement tallgrass prairie. Management of the Preserve incorporates both, though the timing, duration, and recurrence have been somewhat regularized. In contrast to the Kissimmee restoration plan, the Preserve's planners expressed a desire for a management scheme that exhibited variability and unpredictability, to mimic the stochasticity of nature. In this scheme, "in order to allow for the full expression of the tallgrass prairie ecosystem, elements of randomness should be encouraged. The complex interrelationships found within the prairie ecosystem, especially those involving fire and grazing, should be perpetuated in such a way as to ensure that the same activity (such as fire or grazing) does not occur in the same area, in the same way, at the same time, every year (USDI 2000)." Such management would encourage heterogeneity in space and time, as different patches of habitat would be in different stages of succession at any given time. In this way, a random management scheme would maximize the diversity of the overall system.

But heterogeneity and stochasticity are not compatible with the management of a commodity. Rather, predictability is desirable. The rigid management plan that ultimately was adopted for the Preserve is far from random: "The prairie vegetation, under the current grazing lease, is burned every spring, usually around March 20th; The vegetation is subjected to an early intensive stocking regime, averaging two acres for a 550-pound steer for approximately 90–100 days between April 15th and July 31st. The cattle are then removed and the vegetation is allowed a period of regrowth until the next spring" (USDI 2000). So it is true that the native tallgrass prairie is protected here, with controlled burns substituting for wildfire and cattle for bison. But to accommodate humans and their commodities, the disturbance regimes have been regularized and homogenized.

Fifty miles to the north of Tallgrass National Prairie is Konza Prairie, an 8,600 acre Preserve owned jointly by the Nature Conservancy and Kansas State University. A site for both conservation and long term ecological research, Konza has been subjected to a variety of experimental fire regimes for nearly 40 years. In the mid-1980s, a herd of bison was introduced, adding a second major force for the study of disturbance regime on species composition, diversity and productivity, soil and water characteristics, and ecosystem processes (Yaffee and Phillips 1996). Long term ecological research is uncommon; to have sustained research on such a rare ecosystem is valuable indeed.

The grand research design at Konza is to treat different watersheds of the prairie with different fire regimes, including annual fire and a variety of longer return intervals, along with unburned controls. Some sites are mowed, others are grazed. Within this research scheme a great many experiments have been and are being conducted; here I will focus only on two observations. The first comes from 15 years of Konza research on plant, breeding bird, grasshopper, and small mammal communities in areas of different fire regimes (Collins 2000). The biological communities were hypothesized to be most stable when subjected to their historic fire return interval – thought to be 3–5 years. The reasoning is that the historic fire regime maintains the biological community – not necessarily each species in the same relative abundance, but the dominant species, their functional groups, and their structural relationships. More frequent fire, according to the hypothesis, will not allow for the persistence of the dominant tallgrass prairie species, causing the community to be invaded by new species. A much longer return interval was similarly expected to cause a shift in the community, allowing species that were previously excluded by fire to encroach and out-compete the previous dominants.

The results showed that different fire regimes result in different plant communities as expected, but also and more surprisingly that *all* of the plant communities at Konza were undergoing directional change regardless of fire regime (Briggs et al. 2002). Further, the analysis showed that the animal communities of the same sites were all undergoing changes as well – but changes that were unrelated to the plant communities with which they had been associated. In other words, there apparently is no "typical" plant or animal community that exists at the historic fire return interval – or at any return interval. The management implication is profound: the notion of the tallgrass prairie as a stable, characteristic ecosystem is a conceptual oversimplification that does not exist in practice.

A second observation on the Konza experiment shows that human management is often intended to preserve and protect ecosystem characteristics that we find desirable, even in spite of successional change. Portions of the Konza have remained unburned for many years, and on these sites "litter accumulates, woody species invade, moisture and nutrient availability increase, and mesic grasslands eventually develop into shrubland and woodland vegetation" (Collins 2000). Without the human management of controlled burns, then, what would this ecosystem be? Perhaps it would not be a grassland at all. And what if it became something different? Would there be a great loss if a tallgrass prairie became a shrubland? Conversely, is there anything we lose by maintaining these lands as our ancestors found them?

Six Rivers National Forest, California

As its name implies, Six Rivers National Forest is no single ecosystem. Encompassing over one million acres and including 137,000 acres of old growth forest and over 350 miles of wild and scenic rivers, it is a site of both preservation and conservation.

In this regard it is a suitable representation of American national forests in general, which are expected to serve a variety of human needs. Portions of Six Rivers simultaneously serve as ecological sanctuary, recreation area, habitat for threatened and endangered species, timber resource, salmon and steelhead fishery, and wilderness area. It is a vast and impressive resource and a management challenge. But in this brief introduction I will focus on a single species that complicates the preservation and conservation of Six Rivers: *Phytophthora ramorum.*

Phytophthora ramorum is a fungus-like pathogen that was first discovered in California in 1995 and described as a new species in 2000. It infects a dozen or more host plants, but it is particularly lethal to various species and relatives of red oak. The oak, tanoak, and madrone species are substantial components of redwood and mixed evergreen forests of northern California and southwest Oregon. Since its discovery, *P. ramorum* has been blamed for the death of ten of thousands of oaks in and around Six Rivers. It kills with remarkable speed upon infection, giving rise to the name of the disease: Sudden Oak Death (Rizzo and Garbelotto 2003).

The origin of this pathogen is not entirely clear. The species was unknown in the US and Europe prior to the mid-1990s, and the American and European populations appear to be distinct. It may have been introduced into the western US from Europe, or it may have been introduced into both regions from a third location. It is also possible that the species has existed in California for a long time – it may, in fact, be native – and due to some change in its environment or expression has only recently become aggressive and virulent. Whatever its origin, it appears to infect other species, like rhododendron, huckleberry, bay laurel, and California buckeye in a non-lethal way. These associated hosts may facilitate infection of oaks by serving as sites for the production and transmission of spores. In this way, forests with a greater diversity of hosts may be at greater risk of infection (Rizzo and Garbelotto 2003).

Infected trees die from a sort of girdling that restricts nutrient flow through the trunk. First cankers appear on the trunk, surrounded by dead tissue that oozes black sap. Secondary infections of fungi and beetles are common. Once crown dieback begins, the tree is generally lost within a few seasons. The non-lethal infection of other species is known as Ramorum Blight, and is characterized by twig and leaf discoloration and dieback. At present, there is no effective cure or prevention for either the lethal disease or the blight. The primary attempt at prevention is tree removal. Removing California bay laurels near uninfected oak stands may reduce spore stocks in the area and decrease the likelihood of oak infection. But bay laurels themselves are native trees with high wildlife value, and in some areas they may be the best candidates for dominance should oaks be lost. Fungicide has been tested as a preventative measure for healthy oaks; it may prevent the spread, but it is ineffective if the tree has already been infected. Given the enormity of the forest, this measure is potentially useful for only high value landscape and nursery trees (Rizzo and Garbelotto 2003).

This disease and others like it raise important questions for ecosystem management. First, what are the ecological ramifications of the loss of a substantial number of individuals from a dominant species group over a short period of time? Certainly, the dominant oak will be replaced by a new dominant species in patches of extreme mortality. What might this mean for species that thrived under and in the oak canopy? Will they, too, disappear? How closely are species interdependent? And it raises questions about aggregate function. Will the forest ecosystem function differently with the loss of oaks? If so, what functions will change, and how? How about the function of the soil and decomposer ecosystem of the forest floor, or the function of adjacent aquatic ecosystems? Assuming for the sake of argument that the answer to all of the above is *yes* – meaning that the functions of all of these associated systems will change – we may wonder whether the changes will be in any way detrimental to the Six Rivers ecological complex. And most importantly: should humans, as keepers of the national forest, do anything to prevent or slow this change? Or, alternatively, should we let the disease run its course (Rizzo et al. 2005)?

These are questions without easy answers or much concrete evidence, but they are worth considering for their broader application to the practice of ecosystem preservation and conservation. In general terms, they are questions about the origin, mechanism, and outcome of ecological change. First, let's consider origin. As I have noted, the origin of this pathogen has not been determined; it seems likely that it is a nonnative species that was accidentally introduced to California, but it is possible that it could be a native species. Would definitive evidence one way or the other change the way we think about the pathogen and its effects? As a native species, should its action be considered a part of natural ecological succession? If it is determined to be nonnative, should its action then be considered unnatural? It seems likely that, native or not, this pathogen is and will continue to be a threat to this forest and its associated ecosystems. So perhaps species origin does not matter all that much. A related question is this: is the extent and severity of the disease in any way a result of human activity? Sketchy evidence suggests that Sudden Oak Death may be more prevalent in areas that have not burned in the last 50 years and that are, coincidentally, near urban areas. Are trees in these areas subjected to some anthropogenic stress that makes them more susceptible to infection or less able to survive infection? So far, little is known about where this bug came from and why it acts the way it does.

Very well then; by what mechanism is it a threat? Clearly, with unchecked infection, Sudden Oak Death will alter the forest. There has already been and will continue to be substantial loss of valuable timber, but the damage is not only economic. The loss of such dominant trees will change the way the forest looks, though the aesthetic damage may be apparent to only the keenest observer. Of course, successors of deceased oaks may themselves be attractive. How about the ecosystem as a whole? What species or processes will die with the oak? On this topic little is known with certainly, though it seems clear that Sudden Oak Death has influenced some ecological processes already. Tree mortality, for example, may add to the forest fuel load and increase the probability, and perhaps the severity, of wildfire. For humans that live, work, or own property nearby, this is an obvious threat. It could also be regarded as an ecological threat, depending on the intensity and extent of fire and the effect on other species. Another ecological condition that is at risk is the soil surface. The combination of increased organic detritus and exposed soil may change the characteristics critical for seed germination. All told, the dominant trees that succeed the oak may exist within a very different community.

And this leads us to ecological outcomes. What will the forests of Six Rivers – and other forest ecosystems which *P. ramorum* may infect – look like a century from now? It may be that these oaks and tanoak forests eventually become bay laurel dominated systems. Myriad other factors, like climate change, fire frequency, and the emergence of other pathogens make such speculation tenuous at best. The only thing that does seem certain is that the loss of oaks and tanoaks will result in an altered forest ecosystem. Whether this new ecosystem will be better or worse, whether the change should be fought, endured, or celebrated, and whether such change should be regarded as avoidable or inevitable depends upon one's perspective.

Four Questions

In these four examples – oak savanna, river-wetland complex, tallgrass prairie, and multiple-use forest – we see some current efforts to preserve and protect ecosystems. In all four, the management has been planned in considerable detail, not only in terms of what the ecosystem should be but also in terms of what ecological states are to be avoided. Clearly, these are areas that deserve protection. The loss would be great if the oak openings were all converted to suburban housing developments, if the Kissimmee were merely a conveyance for wastewater, if the tallgrass prairies were completely plowed, or if disease and fire consumed our national forests. We have already lost the majority of our ecosystems to such circumstances, making protection of those that remain all the more critical.

In order to preserve these ecosystems, certain types of activities have been excluded, so that the areas are not plowed, cleared, paved or developed. But restriction does not seem to be enough. In all four examples there is the sense that the ecosystem should be as it was historically, as we first encountered it, or in an optimal configuration for ecological or human services. In all four, humans have employed one or more disturbance regimes to manage the ecosystem: fire, flood, grazing, and even mowing, thinning, and herbicides. In all four cases managers have worked to reintroduce desirable organisms and eliminate, or at least minimize, undesirable organisms. In these ways we defend our protected ecosystems.

On a short time scale, an ecosystem may be deemed protected if the land on which is exists has been exempted from human development. We may think of them as preserved or restored ecosystems if species and processes that were present when we first took note remain intact. We may call them sustainable or conserved ecosystems if they are able to consistently provide a commodity or service that is useful to humans. But the protection of ecosystems in a certain desirable state seems to be at odds with the nature of environmental conditions and living things. Ecosystems are dynamic. Species come and go, and physical conditions change.

What, then, should be protected, and how should we protect it? In the United States, over 280 million acres – about 12% of the nation – is under some degree of environmental protection (Vale 2005). An additional three million square miles of aquatic habitat is protected. In this respect the United States is among the world leaders in ecosystem protection. But what is meant, exactly, by protection? Are these ecosystems protected in such a way that they have the capacity for response to disturbance? Are they protected *from* change or *for* change?

In this work I examine the goals and methods of ecosystem protection and their evolution throughout the history of the American conservation movement. In particular, I evaluate the degree to which ecosystem disturbance and spatiotemporal change have been incorporated into the American concept of ecosystem protection. Specifically, I address questions on four issues that have been raised in various ways by the examples in this chapter:

1. Protection of the Ecosystem as a Unit. Knowing, as we do, that species do not occur in communities that are discrete and unchanging, on what basis do we seek to preserve ecosystems as units?
2. Expectation of Stability. With abundant evidence that species and their associations respond to environmental change, and with the knowledge that environmental conditions are rarely constant, why do we expect our protected ecosystem to remain in a stable state?
3. Arrested Succession. Since the ecosystems we desire to protect are the product of succession, which we know to be indeterminate, why do we so often view protection as the maintenance of an ecosystem in a particular successional stage?
4. Disturbance and Response. Recognizing that all ecosystems are subjected to periodic disturbances and that a system's response to disturbance is a function of chance patterns and processes, why do we prize an ecosystem's ability to remain unchanged in the face of disturbance?

All of these questions revolve around the notion of the ecosystem as an entity with an optimal state – with an ideal and enduring form. The American approach to ecosystem management is predicated on the maintenance of the ideal state. It is an approach to management based upon traditional ecological views that are now under assault. What can preservation and conservation mean if ecosystems have no optimal state, no ideal and enduring form? To fully understand the question, we must first explore the roots of American ecosystem protection.

References

Anderson, D., Bousquin, S., Williams, G., and Colangelo D. 2005. Defining success: expectations for restoration of the Kissimmee River. West Palm Beach: South Florida Water Management District.

Brewer, L. and Vankat, J. 2006. Richness and diversity of oak savanna in northwestern Ohio: proximity to possible sources of propagules. The American Midland Naturalist 155:1–10.

Briggs, J., Knapp, A., and Brock, B. 2002. Expansion of woody plants in tallgrass prairie: a fifteen-year study of fire and fire-grazing interactions. The American Midland Naturalist 147:287–294.

Collins, S. 2000. Disturbance frequency and community stability in native tallgrass prairie. The American Naturalist 155:311–325.

De Luise, A. 2006. Florida completes 100,000 acre land acquisition for Kissimmee River restoration. Florida: Florida Department of Environmental Protection Press Release, April 11, 2006.

Goldthwait, R. 1959. Scenes in Ohio during the last ice age. Ohio Journal of Science 59:193–216.

Grundel, R., Pavlovic, N., and Sulzman, C. 1998. Habitat use by the endangered Karner blue butterfly in oak woodlands: the influence of canopy cover. Biological Conservation 85:47–53.

Jones, B. 1999. Archeological overview and assessment for Tallgrass Prairie National Preserve, Chase County, Kansas. Technical Report No. 61. Lincoln: United States Department of the Interior, National Park Service, Midwest Archeological Center.

Koebel, J. 1995. An historical perspective on the Kissimmee River restoration project. Restoration Ecology 3:149–159.

Mayfield, H. 1962. The changing Toledo region: a naturalist's point of view. Northwest Ohio Quarterly 34:83–104.

Pielou, E. 1991. After the Ice Age: The return of life to glaciated North America. Chicago: University of Chicago Press.

Rizzo, D., and Garbelotto, M. 2003. Sudden oak death: endangering California and Oregon forest ecosystems. Frontiers in Ecology and the Environment 1:197–204.

Rizzo, D., Garbelotto, M., and Hansen, E. 2005. Phytophthora ramorum: integrative research and management of an emerging pathogen in California and Oregon forests. Annual Review of Phytopathology 43:309–335.

Sampson, H. 1930. Succession in the swamp forest formation in northern Ohio. Ohio Journal of Science 30:342–348.

Savage, C., Williams, J. A., and Page, J. R. 2004. Prairie: A Natural History. Vancouver: Greystone Books.

Teller, J. T. 1987. Proglacial lakes and the southern margin of the Laurentian ice sheet. In The geology of North America Vol. K-3, North America and adjacent oceans during the last glaciation, ed. Ruddiman, W. F., and Wright, G. E., pp. 39–69. Boulder: Geological Society of America.

Tolson, P., Magdich, M., Seidel, T., Haase, G., and Fazio, B. 1999. Return of a native. Endangered Species 24:14–15.

USDI. 2000. Final general management plan/environmental impact statement, Tallgrass Prairie National Preserve, Kansas. Washington: National Park Service, United States Department of the Interior.

Vale, T. R. 2005. The American wilderness: reflections on nature protection in the United States. Charlottesville: University of Virginia Press.

Warne, A., Toth, L., and White, W. 2000. Drainage-basin–scale geomorphic analysis to determine reference conditions for ecologic restoration—Kissimmee River, Florida. Geological Society of America Bulletin 112:884.

Whalen, P., Toth, L., Koebel, J., and Strayer, P. 2002. Kissimmee River restoration: a case study. Water Science and Technology 45:55–62.

Yaffee, S., and Phillips, A. 1996. Ecosystem management in the United States: an assessment of current experience. Washington: Island Press.

Part I
Ecosystems in Theory

Chapter 2
The Ecosystem Idea and Ideal

American attitudes regarding the protection of natural resources, wilderness, and recreation areas have undergone dramatic changes in the past two centuries, and our current disposition can be understood only in this historical perspective. It is perhaps not surprising that the science of ecology came of age in America at the same time as the conservation movement, but what is surprising is that these two fields have not always informed – or even associated with – one another. Evolutionary ecology in particular has had an uneasy relationship with the policy and practice of ecosystem protection, meaning that the understanding of species assembly and the management of ecosystem succession have not always been in lockstep. This chapter is a sketch of key ideas that have contributed to American ecosystem management in policy and practice.

Preservation, Conservation, and Ecology

The end of the nineteenth century was a time of great popularity for nature in American literature and art, driven in no small way by the works of Catlin, Thoreau, Marsh, Olmsted, and Audubon. For many, the astonishingly rapid and thorough exploitation of American natural resources stimulated recognition of loss and a cry of protest. But even among those who sought to protect natural areas there were conflicting views on the reasons for protection, with some arguing that nature should be protected for its own sake and others envisioning nature reserves that would ensure a continuous stock of commodities for future harvest. One of the best-loved advocates of the former perspective, the preservationist movement, was John Muir (1838–1914). Muir's outlook was initially one of wise use, but he eventually came to the conclusion that the use of a land's commodities, as in lumbering, grazing, damming and mining, were incompatible with its preservation. Increasingly outraged at the wanton destruction of God's creation, Muir came to advocate a level of ecosystem protection that excluded the logger, the rancher, and the developer and left nature in its natural state, to be enjoyed by the hiker, the camper, and the student (Miller 2001).

D. J. Spieles, *Protected Land*, Springer Series on Environmental Management,
DOI 10.1007/978-1-4419-6813-5_2, © Springer Science+Business Media, LLC 2010

The science of ecology, then in its infancy, was not closely associated with the early calls for preservation, but clearly there were scientists thinking about the characteristics of natural areas. Prior to the mid-nineteenth century, the nearest thing to ecology was classification-based natural history, and there was not great attention given to the interaction of species with environmental conditions. Darwin's (1859) seminal work stimulated such an interest. Though Darwin did not brooch what Coleman (1986) calls the "character and action of the bonds that joined organism, community, and environment," his challenge to the notion of the fixed species had implications for change at the community level (Bowler 1993). One of the first to place Darwinian selection within the context of the biological community and its abiotic environment was the Danish scientist Eugenius Warming (1841–1924). "Hitherto," Warming wrote in 1895, "we have treated plant-communities as if they were static entities, in a condition of equilibrium and with their evolution concluded, and were living side by side at peace with one another. Yet such is by no means the condition of affairs" (Warming et al. 1909). Warming connected the struggle for existence within and among species with the environmental conditions in which they struggled. Further, he showed that even a slight change in environmental conditions could alter the complement of species in a community and the species' relationships with one another. In effect, Warming transformed the ecological thought of his day by conceptualizing the biological community as an assemblage "whose coherence is expressed in adaptations, a common manner or form of life, a shared economy" (Coleman 1986). There is a holistic aspect to this early view of the ecosystem, as indicated by Warming's communities "linked and interwoven into one common existence" proceeding through succession toward the "final community." But there is also a notion of species individualism: "each member of a community exists in morphological, anatomical, and physiological agreement with the diverse ecological and social conditions under which it lives," particularly for plant communities, "a congregation of units among which there is no co-operation for the common weal but rather a ceaseless struggle of all against all" (Warming et al. 1909). Warming's recognition of community co-evolution amidst spatiotemporal variation was an early glimpse of ecological paradigms that are still debated today.

Warming's work was followed by that of Frederic Clements (1874–1945), who studied plant communities in the American west and became the most influential American botanist of his time. Warming's concept of the coherent community is apparent in Clements' work, as is Henry Cowles' assessment of ecological succession in the dunes of Lake Michigan (Cowles 1901). Clements saw ecosystem succession as an orderly progression of developmental community stages toward the most advanced level: the mature climax community. In the Clementsian ecosystem, climate and geography set the stage for the development and natural disaster might temporarily interrupt it, but ultimately the community would reach stable maturity. Clements went so far as to consider the ecosystem as a superorganism – a view that was not solely Clements' but in fact a common perception since the time of Plato: "All the stages which precede the climax are stages of growth. They have the same essential relation to the final stable structure of the organism that seedling and growing plant have to the adult individual" (Clements 1916; Kricher 2009). Implicit in the

Clementsian view was the idea that some ecological states are superior to others, and given the opportunity these states will emerge and remain stable.

The influence of Clementsian succession on American ecosystem management can hardly be overstated. According to Pyne (1997), "it was Clements who adapted the European ideas of ecology to the American landscape. It was the Clementsian concept of ecology that entered American forestry and land management." It is an ecosystem concept that remains influential today – in fact, classic Clementsian succession as advanced by the work of E.P. Odum is commonly portrayed in modern textbooks (Fig. 2.1; See commentary by Gibson 1996). A few aspects of Clements-derived management bear note, for we will see them again. In Clementsian ecology, disturbances such as fire, flood, or drought are seen as hindrances along the path to the climax.

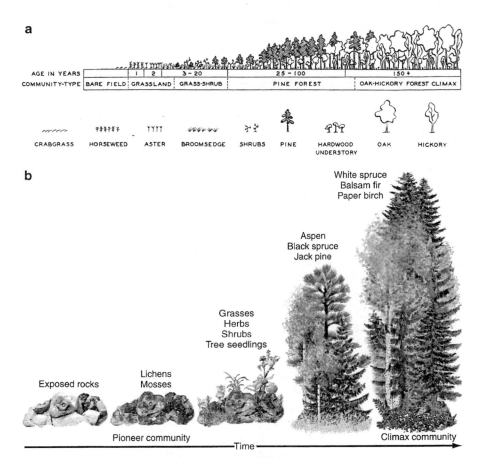

Fig. 2.1 Classic Clementsian succession as depicted (**a**) by E. P. Odum (1956) and (**b**) in a modern textbook (Cunningham and Cunningham 2009). *Top* diagram republished with permission of the Ecological Society of America from Breeding bird populations in relation to plant succession on the piedmont of Georgia by Johnston and Odum 1956, Ecology 37(1):51; permission conveyed through Copyright Clearance Center, Inc. *Bottom* diagram reprinted with permission of the McGraw-Hill Companies, Inc.

One may achieve a climax state in an ecosystem by managing or preventing the disturbance. Absent these interruptions, a community will move through stages in a deterministic manner, with the biota of each stage making the environment suitable for the next stage. The mature climax stage reaches a point of stabilization that is self-perpetuating, to be undone only by disturbance. And, finally, a notion that was assigned perhaps unfairly to Clements: that succession progresses teleologically toward an ultimate purpose (Clements 1916; Hagen 1988).

One could imagine that the Clementsian view might fit nicely with John Muir's desire for preservation. Both envision an ideal state for a given ecosystem. In both, there is the idea that an ecosystem, left undisturbed, will achieve its proper state. Muir was intimately familiar with Darwinian evolution but had difficulty with the idea of a random, purposeless nature. He could not accept Darwin's view of the brutality of nature; Muir consistently considered even the destructive events in nature to be essentially benign, kindly, and harmonious (Fox 1985; Wilkins 1995). He was truly a holist who thought of ecosystems as an organic unit. Though he was aware of mechanisms of ecosystem change, the "natural state" to Muir was the pristine ecosystem, unspoiled by humans, and it was this state that he sought to protect. "God has cared for these trees, saved them from drought, disease, avalanches, and a thousand straining, leveling tempests and floods; but he cannot save them from fools – only Uncle Sam can do that" (Muir 1901).

Clements' work was equally compatible with utilitarian conservationism, and it ironically became ammunition for one of Muir's rivals. Gifford Pinchot (1865–1946) was appointed as chief of the Division of Forestry in the Department of Agriculture in 1898 (Miller 2001). In 1905, Pinchot engineered the transfer of federally-owned forests from the Department of the Interior to his jurisdiction in the Department of Agriculture. This was representative of a national change in direction toward management and regulation of federally protected land, supported by President Roosevelt and administered by Pinchot. Pinchot had had virtually no education in the general life sciences when he became the chief forester of the country, and only 1 year of forestry training in France. His approach to forest management was a blend of the European philosophy of uniform geography and highly managed thinning, harvest and regeneration in successive cycles, and the influence of his mentor Frederick Law Olmsted (1822–1903), who promoted a landscape architecture meant to portray the beauty of nature.

Olmsted shaped Pinchot's vision of forestry and, by extension, a century of American ecosystem conservation (Roper 1973). Inspired by European models of ecosystem management, Olmsted's landscape design conformed to the existing topography and features of the land. Attractive scenery, striking vistas, and subtle effects of outcroppings, meanders, and hummocks are combined in Olmsted's work to make the picturesque appear spontaneous. Picturesque it was, but also contrived. Olmsted's design – often for urban parklands – was one of planned placement and context. This approach to landscape architecture was conveyed from Olmsted to Pinchot, whom Olmsted recommended to manage the Biltmore Forest of the George Vanderbilt estate in North Carolina. It was Pinchot's first big opportunity as a forester, and with Olmsted's consultation Pinchot devised plans of selection, harvest, extraction, and regeneration – even down to the preservation of certain trees "for effect" (Pinchot and Steen 2001). Pinchot's forestry was clearly a different endeavor than Olmsted's parkland architecture, but it bore the same stamp of the planned ecosystem, the landscape of purpose.

Pinchot's ecological view seems to have been a mixture of Darwinian natural selection, to which he clearly had been exposed, Olmstedian purposeful design, and the Clementsian concept of the climax community. In his *Primer of Forestry* (1900) Pinchot offers a Darwinian description of the trees that are targeted for harvest:

> Natural selection has made it clear that these are the best trees for the place. These are also the trees which bear the seed whence the younger generations spring. Their offspring will inherit their fitness to a greater or less degree, and in their turn will be subjected to the same rigorous test, by which only the best are allowed to reach maturity. Under this sifting out of the weak and the unfit, our native trees have been prepared through thousands of generations to meet the conditions under which they must live.

But in the same document he insinuates the climax community as an endpoint:

> The trees of the mature primeval forest live on, if no accidents intervene, almost at peace among themselves. At length all conflict between them ends.

The "accidents" to which he refers are disturbances. For Pinchot, they included human-caused events, such as lumbering and grazing, and particularly natural events like fire, wind, and pest infestations. Pinchot saw natural disturbances as a great waste, and an avoidable waste at that. He advocated prevention of these events where possible, and above all the management of nature. He had the law on his side; the federal Forest Management Acts of 1891 and 1897 were intended to "regulate use of and preserve forests from destruction" (Miller 2001). Protected from destruction and allowed to regenerate after harvest, Pinchot's forest would in time return to the climax and again be ready for harvest. Clementsian succession thus allowed for exploitation of an ecosystem, for the climax could regenerate itself, and would if given the chance.

Pinchot and Muir have at times been portrayed as polar opposites of the environmental movement, but in fact both played a critical role in the nation's recognition of natural resources and their need for some level of protection. Muir's life and work have inspired generations of environmental activists and advocates for the preservation of wilderness. Pinchot, for his part, was instrumental in doubling the number of national parks to ten, set aside 18 national monuments, and established more than 50 national bird sanctuaries (Miller 2001). Late in his career, Pinchot's utilitarian forestry even softened a bit as he acknowledged – at the urging of Clements – the need for a more ecological approach to forestry, including an understanding of plant diversity, soils, insect community dynamics, and "the balance of nature" (Pinchot 1937). Clearly, both Muir and Pinchot prevented the privatization of American forests from doing even greater harm than it has done. But my purpose is not to debate their relative contributions to American society. Instead, it is to recognize that Muir's preservation and Pinchot's conservation were both based upon Clementsian reasoning and as such both arrive at the same place: protection of the ecosystem in its ideal form. In forestry and wilderness advocacy, this notion shaped American ecosystem management for decades.

Gleason and Individualism

Pinchot's late acknowledgment of ecological systems notwithstanding, the conservation movement was informed by ecological research in only a limited way in the early twentieth century. Increasingly, though, ecologists were challenging the Clementsian view of the successional superorganism. Henry Gleason (1882–1975) tramped over much of the midwestern United States analyzing plant communities and came to the conclusion that a plant community was not a developmental stage of some inevitable climax. Rather, he suggested, a plant community was the result of two criteria: (1) the local conditions, such as soil, topography and climate, and (2) the plants that happened to be available for colonization (Nicolson 1990). Further, Gleason noted profound differences in plant communities of the beech-maple forest from Lake Superior to the Ohio River, leading to the conclusion that there was no typical beech-maple community. This was an individualistic ecology, with successional variation in space and time, and it flew in the face of Clementsianism. Clements enjoyed wide support at the time, and Gleason later noted that "for 10 years, or thereabout, I was an ecological outlaw" (Gleason 1987).

As others advanced the state of field ecology, the Clementsian view was increasingly met with skepticism, though the models of Clements and Gleason were by no means the only two in existence (Nicolson 1990). British ecologists had never fully accepted the idea of an ideal climax state; the botanist Arthur Tansley and the animal ecologist C.S. Elton were both outspoken critics of Clementsian model. Elton rejected the idea of the "balance of nature," arguing instead that populations of species fluctuated continually and unpredictably in the face of selection pressures. Tansley (1935) found the Clementsian successional units to be "nothing but the synthesized actions of the components in association." But Gleason became the American face of individualistic ecology, and he challenged the Clementsian view on at least four counts: (1) that plant communities occurred in typical associations; (2) that these associations were developmental stages progressing toward a particular climax community; (3) that, in a given climate, the successional progression was determined by the plants themselves; (4) that the developmental associations and climax stages were held together as units. Gleason's model harkened to Warming's individualism; it included the idea that there are physiological differences between species and that plant colonization and hence plant association characteristics vary along environmental gradients. A species' presence in a community, then, was more a matter of chance than design: "Are we not justified in coming to the general conclusion…that an association is not an organism, scarcely even a vegetational unit, but merely a co-incidence?" (Gleason 1926). Tansley put it another way in 1929, stating that a climax community "is a mere aggregation of plants on some of whose qualities as an aggregation we find it useful to insist" (Golley 1996). The Clements-Gleason debate raged into and beyond the 1930s.

Leopoldian Preservation and Conservation

The intellectual discordance of Muir's preservation, Pinchot's conservation, Clements' orderly succession, and Gleason's individualistic ecology was apparent in the career of Aldo Leopold (1887–1948). A graduate of the Yale School of Forestry, Leopold's early career was clearly influenced by Pinchot. For instance, as part of a management plan to increase the number of deer for hunters in the American southwest, the young Leopold advocated the extermination of the wolves and mountain lions that held the deer in check – a policy that he later regarded with regret (Strong 1988). Widely acknowledged as a champion of wilderness protection, Leopold also understood the human and ecological need for nature's commodities, for "who knows for what purpose cranes and condors, otters and grizzlies, may some day be used?" (Leopold and Schwartz 1966). But he questioned absolute anthropocentric conservationism, wondering "whether the principle of highest use does not itself demand that representative portions of some forests be preserved as wilderness" (Leopold 1921). In this respect he emulated much of Muir's philosophy of nature preservation and nature study for the benefit of the human spirit; "raw wilderness gives definition and meaning to the human enterprise" (Leopold and Schwartz 1966). Leopold the scientist saw ecosystems as energy and material flowing through a biotic community, and early in his career he understood community succession in the Clementsian concept. In writing about the brushlands of southern Arizona, Leopold noted that "the climax type is and always has been woodland...this transition type is now reverting the to the climax type" (Leopold 1924; Meine 1988). The climax was "a base-datum of normality, a picture of how healthy land maintains itself as an organism" (Leopold 1941).

Leopold embodied the ideal state of an ecosystem in his call for ecosystem health, integrity, and stability – concepts that would become central to the ecosystem approach adopted by American land management agencies some five decades later (Grumbine 1998). During Leopold's career, the U.S. Forest Service began an effort to protect some of the nation's forests as wilderness areas. Initially, protection as wilderness was still much in the spirit of Pinchot's conservation, meaning that the reserves were designated for protection temporarily, until a future use was determined (Vale 2005). Robert Marshall, director of the U.S. Forest Service Division of Recreation and Lands, issued more stringent protection in 1930, including permanent prohibitions on timbering and road building on land designated as wilderness. A co-founder of the Wilderness Society, Marshall's ecology was also based on the ideal successional state as it occurred when white men first laid eyes on it. His motives were to protect the ecosystem in its climax, for "a wilderness without developments for fire protection will sooner or later go up in smoke and down in ashes" (Marshall 1930).

In the wilderness preservation of Leopold and Marshall, one sees the legacy of John Muir – seeking to protect nature so that it might achieve its intended state, "as museum pieces, for the edification those who may one day wish to see, feel, or study the origins of their cultural inheritance" (Leopold and Schwartz 1966). Indeed, this was in keeping with the National Park Service Act of 1916 which called for conservation of scenery, natural and historic objects, and wildlife in such a way that they be left "unimpaired for the enjoyment of future generations" (Strong 1988). For the young National Park Service, "unimpaired" referred more to scenery and the opportunity for tourism and recreation than to ecological state, but the effect was unquestionably ecological. Beginning with the tenure of Stephen Mather, who served as the first National Park Service Director from 1917 to 1929, the NPS managed the parks as though they were maintaining ideal climax communities. Over the protests of a small staff of wildlife biologists, destructive events like fire and predation were minimized or eliminated, popular wildlife was encouraged for viewing, fish were raised in hatcheries and stocked for sport fishing, and exotic organisms were eliminated when their presence was undesirable – but introduced when they were deemed useful, as in the case of sport fish (Sellars 1997).

The effort to appeal to tourists was extraordinarily successful, and the number of visitors to the national parks skyrocketed in the 1950s. This prompted an effort to develop those areas of the park most visited with roads, lodging, water and sewer facilities, parking lots, trails, and buildings – all of which perpetuated the perceived need to preserve those areas in the state that was compatible with infrastructure and in keeping with visitor expectations. In this way, the NPS was truly attempting preservation, just as American conservation efforts at the time attempted to perpetuate forest resources for future harvest. But Gleason's challenge to the "ideal state" of an ecosystem, either for preservation or conservation, would eventually change the way the scientific community considered ecosystems and would illustrate deficiencies in the conservation and preservation mindset of Leopold, Marshall, and Mather.

Two decades after Gleason published on individualistic ecology, a young graduate student at the University of Illinois waded into the debate on ecological succession. Robert Whittaker's (1920–1980) dissertation tested the idea of a stable, co-adapted community by investigating plant communities along an elevation gradient in the Smokey Mountains. In one 4-month field season, he gathered data from 300 random locations. What he found was a great bit of empirical support for the Gleasonian view. "My hypothesized groups of co-adapted species with parallel distributions were not there and the transitions I had been looking for were not in evidence as such, since the many kinds of forest communities intergraded continuously" (Jensen and Salisbury 1972). Throughout his career, Whittaker wrestled with Clements' idea of the climax based on regional climate, the discrete vegetation associations leading to the climax, and the concept of the superorganism (Westman and Peet 1982). Whittaker was one of a growing number of field ecologists lending evidence to the question, and support for Clementsian succession began to waver. A tremendous blow was dealt by paleoecologists in the 1960s. Margaret Davis, Edward Cushing, and Donald Whitehead analyzed pollen deposits in sediments to reconstruct the geographic distribution of plants in North America since the last glaciation and independently concluded that species have migrated and continue to migrate individualistically, not in predictable associations (Wright and Frey 1965). Viewed over geologic time, Clements' climax community simply does not exist.

Hutchinson, Holism, and Individualism

Evidence against the Clementsian climax did not signal its demise, nor did it mean that ecologists united in one school of thought regarding the nature of ecosystems. No single person represents the divergence of thought in modern ecosystem ecology better than the eclectic and prolific G.E. Hutchinson (1903–1991). An authority on many aspects of ecology, Hutchinson advanced two concepts that are particularly relevant to the current discussion. First, with colleague Raymond Lindeman, he advanced the concept of mathematical modeling in ecology (Lindeman 1942). In this approach, the ecosystem is considered as a defined set of interacting components, such as trophic levels, through which energy, material, and information flow. Modeling enabled the system to be studied and simulated holistically. This was instrumental in giving rise to the field of systems ecology, in which one considers the properties of the ecosystem as a unit. Hutchinson advanced the idea that ecological systems exhibit characteristics of self-regulation that maintained equilibrium conditions. Hutchinson's students, particularly Howard Odum and Robert MacArthur and other ecologists like Eugene P. Odum developed important aspects of community and systems ecology under Hutchinson's influence. The Odum brothers promoted two of the most powerful and long-lasting ecosystem analogies: the ecosystem as a unit in physiological homeostasis and the ecosystem as a self-regulating machine. These images of the classic holistic view became cemented into the American concept of ecosystem management primarily by the Odum book *Fundamentals of Ecology*, the first of its kind and most influential ecosystem text of the next several decades (Odum 1953; Hagen 1992; Golley 1996). Such were the far-reaching effects of Hutchinson the holist: "The evolution of biological communities, though each species appears to fend for itself alone, produces integrated aggregates which increase in stability" (Hutchinson 1959).

At the same time, Hutchinson was an innovator in the fields of population and community ecology. He advanced the concepts of the individual niche and patch dynamics, which effectively constitute a viable alternative to the Clementsian climax (Watt 1947; Hutchinson 1957). In this view, disturbances continually create open space which allows some species to thrive – species that in equilibrium conditions would likely have been out-competed by later successional species. Thus some species exists as "fugitives," hopping from patch to newly-disturbed patch. This means that the complement of species in an ecosystem at any given time is largely dependent upon disturbance and colonization. An ecosystem, in this context, may be thought of as a patterned mosaic, with species shifting independently in space and time.

These two Hutchinsonian concepts – the stable aggregate and the individualistic niche – are not mutually exclusive, but they have come to represent different schools of thought. To illustrate the difference, let us consider how ecological disturbance might be viewed in each. In the holistic view, the ecosystem is considered as a unit which, in its climax, is at equilibrium – meaning that all processes in the system are counterbalanced by other processes, and the system as a whole remains stable (Wu and Loucks 1995). The stable state exists in relation to a state of disturbance, caused by some perturbation. Disturbance events, then, are seen as processes that cause a departure from the stable state. An ecosystem may retain its stable equilibrium in the face of minor disturbance, or, if the disturbance is severe, it may exhibit resilience – the ability to return to the stable state after the event. At its most extreme, the holistic view places great emphasis on the ecosystem as a unit and on stable equilibrium as the default state of that unit. There is also emphasis on the self-regulating, homeostatic capacity of the system via internal feedbacks that tend to maintain the steady state. Modern champions of the holistic equilibrium stop short of the superorganism analogy, but not by much; they see the ecosystem as an object progressing toward a maturity that is self-preserving and resilient in the face of disturbances (Margalef 1963). The system is seen as greater than the sum of its parts; it is a unit which has self-perpetuating emergent properties. To some, this smacks of Clementsianism, but its defenders see it as a mechanistic consideration of the aggregation and interaction of species as they react to and manipulate their environment.

There is another, more diluted concept of holism. To be consistent with the literature I will refer to this as the holistic view in its "weak" form. In this interpretation an ecosystem is comprised of many parts and processes, and none of these components can be understood in isolation. Species and processes, then, necessarily affect and are affected by other species and processes. That is all. There is no implication that the various organisms exist as permanent cogs in a greater machine or that the system is a self-maintaining unit. Are there properties of the system that cannot be accounted for by any of the parts except in their interaction? Absolutely. But it is quite another thing to say that the collection of organisms work as a unit toward self preservation. Weak holism, then, suggests that the system is not greater than the sum of its parts – it is *precisely* the sum of its parts.

Criticisms of the "strong" holistic view are numerous, but in general they are based upon the observations that the "unit" of the ecosystem is an arbitrary abstraction that does not really exist. The emphasis here is on stochasticity of environmental conditions, the patchiness of habitat, and the individualistic and fluid niche of species. In the extreme individualistic view, there is no climax community, and succession is "understood solely in terms of the interaction of individual evolutionary strategies" (Pickett 1976). Accordingly, the idea of a predominant stable state of equilibrium is also questioned, with some advocating a nonequilibrium environment in a constant state of flux, or else multiple states of equilibrium that can be conceptualized only in the context of environmental variation (Holling 1973; White 1979). Environmental variation is patchy in space and time, meaning that an ecosystem is really an observer-defined collection of species that happen to occur together because environmental conditions and historic events favor their congregation at the defined moment. Some have gone so far as to declare the ecosystem – at least the stable, self-regulating unit at equilibrium – to be a myth (Soulé and Lease 1995; O'Neill 2001). Kapustka and Landis (1998) assert that "no human has ever seen an ecosystem" because it is not a discrete unit like an individual organism or even a population.

The New Ecology

"Wherever we seek to find constancy we discover change." Thus wrote Daniel Botkin in 1990, as he among others called for a new approach to ecology and ecosystem management. Actually, several schools of ecological thought have at one time or another been referred to as the "New" ecology, ranging from Eugene Odum's description of systems ecology in 1964 to the even newer perspective of Sven Jorgensen and others in 2007 (Odum 1964; Jørgensen et al. 2007). In essence, Botkin's "new ecology" is nonequilibrium ecology: a recognition of ecosystems as open, complex and dynamic systems that are characteristically transient and unstable. It is a view of nature that does not does not preclude mature populations in stable equilibrium, but holds that ecological communities subjected as they are to chaotic environmental fluctuation rarely achieve such equilibrium states. Rather, nonequilibrium – a condition of inconstancy – appears to be the norm in most ecosystems most of the time. This "nature of change" (Botkin's term) is the result of different mechanisms at different scales (Botkin 1990; Rohde 2005). At the population level, ecological disturbances alter biotic or abiotic conditions of physiological stress and prevent the stable equilibrium; at the community level, disturbances create and re-create new niches for colonization; on the landscape scale mass migrations, extinctions, and catastrophic events continually create new opportunities for novel assemblages. Across spatial and temporal scales, the basic concept is that disturbance and stress drive environmental heterogeneity, preventing successional patches from ever achieving equilibrium (Levin and Paine 1974).

I'll hazard to suggest that nonequilibrium ecology is widely accepted among modern ecologists, partially because of abundant evidence and partially because of the lack of empirical support for the alternative. The point has been driven home by others (Zimmerer 1994; Wu and Loucks 1995; Rohde 2005; Kricher 2009); here I will merely summarize some key points in its favor:

1. Wildlife populations have not routinely been found to fluctuate in a regular manner about some stable value. On the contrary, many population studies suggest that chaotic, instable fluctuation is more the rule than the exception.
2. Evidence for cyclic repetition of typical, homogenous communities in a given environment is lacking. On the contrary, there is abundant evidence that ecological communities at all scales are as spatially heterogeneous as the environment they inhabit and as irregular as chance opportunities for organismal response.

3. Paleoecological evidence has made it abundantly clear that ecological communities are temporally transitional and that the assemblages we see today are but the latest manifestation in a continuum of temporary permutations. Temporal heterogeneity is a property of ecosystems at all spatial scales.
4. Empirical evidence that ecological systems are homeostatic entities is lacking. On the contrary, there is abundant evidence that spatial heterogeneity, stochastic perturbation, and historical contingency are critical factors for the individualistic occurrence and behavior that we see as ecosystem composition, structure, and function.
5. Evidence that ecological communities are self-directed through succession toward an optimal configuration is lacking. On the contrary, abundant evidence suggests that species in loose and ephemeral association respond individualistically to stress and disturbance. Succession, then, is not an orderly progression of increasingly superior associations, but rather a continuously changing patchwork of opportunistic species.

Much has been written about nonequilibrium ecology in recent years, but it is not my purpose here to present a complete review. I offer only two comments. First, this "new" ecology is hardly new; its precepts can be seen in the work of Warming, Gleason, Whittaker, and Hutchinson. Its application to ecosystem management, however, is still new, for holism has long reigned over American ecosystem conservation and preservation. This brings up the second note: nonequilibrium ecology is inconsistent with traditional ecosystem management in the United States. It challenges the holistic ecosystem view as a recapitulation of the Clementsian superorganism, as a deterministic model of some human-defined stable state. Instead, it suggests that environmental variation and chance opportunism regulate ecosystem assembly (Zimmerer 1994).

In their extremes, strong holism and individualistic nonequilibrium represent ends of the conceptual ecosystem continuum (Fig. 2.2). The diametric opposition makes it difficult to approach the idea of protecting an ecosystem; indeed, it even makes the definition of an ecosystem unclear. Consider H.T. Odum's (1994) definition: "an ecosystem is an organized system of land, water, mineral cycles, living organisms, and their *programmatic behavioral control mechanisms* (emphasis added)." A different definition is offered by Wu and Loucks (1995): "ecological systems can be seen as hierarchical systems of patches that differ in size, shape, and

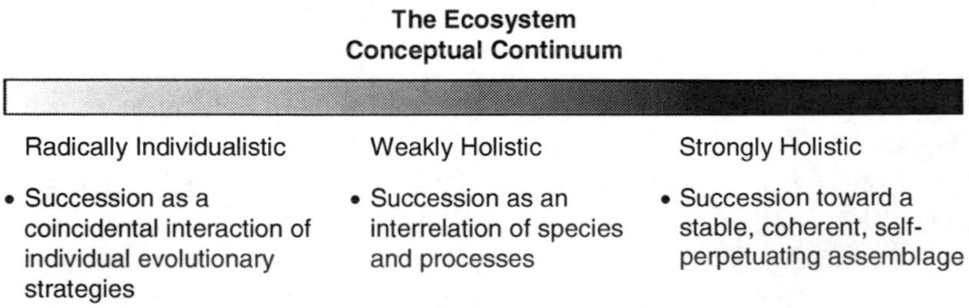

**The Ecosystem
Conceptual Continuum**

Radically Individualistic	Weakly Holistic	Strongly Holistic
• Succession as a coincidental interaction of individual evolutionary strategies	• Succession as an interrelation of species and processes	• Succession toward a stable, coherent, self-perpetuating assemblage

Fig. 2.2 The ecosystem: a conceptual continuum

successional stage at particular scales." To Simon Levin (1992), an ecosystem "is really just an arbitrary subdivision of a continuous gradation of local species assemblages." What does it mean for the practice of ecosystem preservation or conservation to have such a discrepancy of definition? Put another way, if there is no such thing as a stable, optimal state of an ecosystem, what exactly are we protecting? This is a dilemma with which our governmental agencies and conservation organizations continue to struggle.

Preservation of the Ideal

American conservation agencies maintained a holistic perspective through the mid-twentieth century, as indicated by policies of sustained yield in the Forest and Fish and Wildlife Services and by the practice of preserving the natural character of desirable native organisms in the case of the National Park Service. The evolution of the NPS serves as an example of efforts to incorporate science into management and, even more basically, the question of which science to incorporate.

The influence of ecological science was minimized from the inception of the NPS through the 1950s. As we have seen, the overarching emphasis during those early decades was on the preservation of scenery and the development of infrastructure to facilitate tourism. The environmental movement of the 1960s and 1970s challenged the NPS approach to ecosystem conservation and urged the inclusion of science. For example, the Leopold Report (1963), written by a committee headed by Aldo Leopold's son A. Starker Leopold, urged a more active and scientific approach to ecosystem management in the national park system. The Leopold Report promoted an emphasis on wildlife biology and ecological succession:

> Habitat is not a fixed or stable entity that can be set aside and preserved behind a fence, like a cliff dwelling or a petrified tree. Biotic communities change through natural stages of succession.

But it still amounted to preservation of the ideal:

> The goal of managing the national parks and monuments should be to preserve, or where necessary to recreate, the ecologic scene as viewed by the first European visitors. As part of this scene, native species of wild animals should be present in maximum variety and reasonable abundance.

> There is no need for active modification to maintain large examples of the relatively stable "climax" communities which under protection perpetuate themselves indefinitely... However, most biotic communities are in a constant state of change due to natural or man-caused processes of ecological succession. In these "successional" communities it is necessary to manage the habitat to achieve or stabilize it at a desired stage.

The Leopold Report advocated maintaining the desired stage with intensive management, with fire, predation, animal introduction and extirpation, even with earth-moving equipment in the effort to guide succession to a desired state. It urged the inclusion of a decidedly holistic ecosystem science.

In the same year, a National Academy of Sciences Committee released a report on the National Parks – the so-called Robbins Report (Robbins et al. 1963) – that expanded upon many points of the Leopold report. One major thrust of the Robbins report was to call for increased and sustained scientific research in the NPS. On the topic of ecosystem change, the authors seem caught between the acknowledgement that ecosystems are ever-changing and the desire to preserve their "unique features":

> The Committee recognizes that national parks are not pictures on the wall; they are not museum exhibits in glass cases; they are dynamic biological complexes with self-generating changes. To attempt to maintain them in any fixed condition, past, present, or future, would not only be futile but contrary to nature. Each park should be regarded as a system of interrelated plants, animals and habitat (an ecosystem) in which evolutionary processes will occur under such human control and guidance as seems necessary to preserve its unique features. Naturalness, the avoidance of artificiality, should be the rule.

These reports, and increasing pressure from the environmental movement and conservation organizations, resulted in an acknowledgment that a policy change and a greater focus on ecology were necessary in the NPS. Accordingly, an NPS policy statement in 1970 included: "The concept of preservation of a total environment, as compared with the protection of an individual feature or species, is a distinguishing feature of national park management" (USDI 1970). Change was slow, for the NPS was not set up for scientifically-based management decisions. Eventually, research plans and then resource management plans began to be developed for individual parks, but science was still underfunded and generally not part of park administration or management (Sellars 1997). The struggle for the inclusion of science in the upper levels of decision making in the NPS is ongoing.

In 1964 the United States established a National Wilderness Preservation System for the purpose of "preserving and protecting lands in their natural condition." The Wilderness Act of 1964 protected natural areas in a way that differed in concept and purpose from the protection of national parks. The Act defines wilderness in relation to the human: "A wilderness...is hereby recognized as an area where the earth and its community of life are untrammeled by man, where man himself is a visitor who does not remain" (Public Law 88-577). The implication is that succession would be allowed to happen in designated wilderness areas, regardless of outcome. The system was designed as an overlay of existing agency structure, so that wilderness could be designated on public lands managed by various agencies. This was seen as a victory for conservationists who were dissatisfied with the NPS focus on tourism and recreation. In 1974, The Eastern Wilderness Act extended the wilderness preservation effort beyond western lands. Together with new agency policies, The Wilderness Acts represented an acknowledgment that ecosystem protection might entail more than maintenance of the static equilibrium.

Toward an Ecosystem Approach to Management

The environmental explosion of the 1970s brought great increases in environmental legislation and in the number of protected areas. The influence of the holistic eco-system concept was evident in the environmental movement, but there was also a rising tide of evolutionary ecology that provided evidence for the individualistic, nonequlibrium ecosystem, spurred by increasing influence of molecular biology on evolutionary studies (Hagen 1992). The dichotomy of the holistic and individualistic concepts was deemed a simple matter of perspective by some, but for others the two approaches were incompatible. Which perspective won the day?

It would be hard to argue that the ecosystem protection efforts of the 1970s were based on individualistic ecology. The Endangered Species Act of 1973, with its emphasis on protection of critical habitat for rare species, implies the preservation of an ideal state for species on the list. The Clean Water Act and subsequent no-net-loss policy for wetland mitigation treats ecosystems as units. The multiple-use, sustained-yield principles of the National Forest Management Act of 1976 essentially reaffirmed the management policies of Pinchot. The Federal Land Policy and Management Act of 1976 stressed the maintenance of public lands "and their various resource values so that they are utilized in the combination that will best meet the present and future needs of the American people." All of these policies are based on the holistic concept of ecosystems – that humans can and should maintain them in their ideal state. Gleason was no longer an ecological outlaw, but in the American environmental movement Clements still had the upper hand. In the words of Simberloff (1980), "Clements' superor-ganism is not dead but rather transmogrified into a belief that holistic study of ecosystems is the proper course for ecology."

In addition to an avalanche of legislation, the environmental movement ignited a rapid increase in the amount of land under protection in the United States. Between 1970 and 1990 the number of acres in the national parks, wildlife refuges, wilderness areas and forests grew from about 20 million to nearly 200 million acres (Vale 2005). As the amount of protected land grew so too did concerns about the manner of protection and management. In response, the Clinton administration initiated an effort to institute an "ecosystem approach to management" in the federal agencies. In part, the approach was a call for integration of social, eco-nomic, and ecological needs, the collaboration of multiple stakeholders, and the coordination and communication of agencies in the management process. Ecologically, the new approach was intended to promote the "attractiveness" as well as "the health, productivity, and biological diversity of ecosystems…and their functions and values." The approach called for management of an ecosystem toward "a desired future state – the ideal state toward which efforts are directed." At the same time, the approach recognizes that "ecosystems…are complex, dynamic, characteristically heterogeneous over space and time, and constantly changing" (Interagency Ecosystem Management Task Force 1995). Given the dynamic and ephemeral heterogeneity implied by this last sentiment, what can terms like "health," "attractiveness," and "ideal state" mean? Can an ideal ecological state be achieved? If achieved, can it be maintained? Should it be maintained?

During the same period, 1970–1990, the number of acres protected by private conservation easement quadrupled in the United States to about 400,000 acres, and the growth absolutely skyrocketed thereafter, so that by 2005 it was well over six million acres. Many of these easements are held by environmental organizations such as The Nature Conservancy, The Trust for Public Land, Ducks Unlimited, American Farmland Trust and The Conservation Fund, though a significant portion are controlled by local or regional land trusts. Naturally, these organizations are not held to the ecosystem approach as espoused by federal agencies, but conservation easements have their own ecological idealism. Most easements are written to protect scenic or ecologically significant habitat *in perpetuity*. A legal agreement to protect an ecological feature of the land in perpetuity assumes that the ecological feature exists in perpetuity. Such an agreement could be problematic as organisms migrate, boundaries shift, and abiotic conditions change. In this way much private land conservation is based upon the holistic concept of the stable equilibrium (Greene 2005; McLaughlin 2005; Kiesecker et al. 2007).

Consider what we desire to protect. In the case of national parks, the emphasis is on protecting scenery and the opportunity for recreation and tourism. Attractive, aesthetically pleasing areas are seen as best suited for scenery and tourism. In national forests, fisheries, and wildlife refuges we wish to sustain the capacity to harvest desired commodities. The protection of critical habitat for rare species is the impetus for protection of many public and private natural areas. In addition, we desire natural areas that are healthy and biologically diverse. We want our protected ecosystems to be systems of integrity and to remain in their optimal configuration. These sentiments may have noble roots, but scenery preservation, sustainable harvest, maintenance of critical habitat, and perpetuation of individual species are concepts at odds with Gleasonian ecology. They require the "arresting" of a natural area in a particular state, with a particular set of characteristics, or else the rapid post-disturbance or post-harvest return of an ecosystem to its desired state. And yet, it is clear that ecological systems are based on change. Species arrive and depart with virtual independence, disturbances change environmental conditions at multiple scales and time intervals, gradients shift over time, and succession cycles in a patchy, hierarchical mosaic toward no particular destination. In the absence of such change, can a protected ecosystem be "healthy"?

The holistic perspective need not be deterministic or climax-oriented; as an abstraction it can be a useful way to analyze ecological interrelationships. It is also clear that ecosystems can and do have their own properties and perform certain services, even though the system is ephemeral in space and time. However, the utility of the "strong" holistic perspective is limited in real-world ecological protection and ecosystem management, precisely because the ideal state is an abstraction. Even so, American ecosystem protection is largely based upon the holistic abstraction: we work to make our ecosystems healthy, to have integrity, to be stable and resilient. In short, we manage our ecosystems as though they are units that can resist change, rather than coincidental assemblages in the midst of change.

References

Botkin, D. B. 1990. Discordant Harmonies: A New Ecology for the Twenty-first Century. New York: Oxford University Press.

Bowler, P. J. 1993. The Norton History of the Environmental Sciences. New York: W. W. Norton.

Clements, F. E. 1916. Plant Succession: An Analysis of the Development of Vegetation. Washington: Carnegie Institution.

Coleman, W. 1986. Evolution into ecology? The strategy of Warming's ecological plant geography. Journal of the History of Biology 19:181–196.

Cowles, H. 1901. The physiographic ecology of Chicago and vicinity: a study of the origin, development, and classification of plant societies (concluded). Botanical Gazette 31:145–182.

Cunningham, W. P., and Cunningham, M. A. 2009. Principles of Environmental Science, Inquiry and Applications. New York: McGraw-Hill.

Darwin, C. 1859. On the Origin of Species by Means of Natural Selection, or the Preservation of Favoured Races in the Struggle for Life. New York: Appleton.

Fox, S. R. 1985. The American Conservation Movement: John Muir and his Legacy. Madison: University of Wisconsin Press.

Gibson, D. J. 1996. Textbook misconceptions: the climax concept of succession. The American Biology Teacher 58(3):135–140.

Gleason, H. 1926. The individualistic concept of the plant association. Torrey Botanic Club Bulletin 53:7–26.

Gleason, H. 1987. A letter from Dr. Gleason. Brittonia 39:205–209.

Golley, F. 1996. A History of the Ecosystem Concept in Ecology: More than the Sum of the Parts. New Haven: Yale University Press.

Greene, D. 2005. Dynamic conservation easements: facing the problem of perpetuity in land conservation. Seattle University Law Review 28:883–923.

Grumbine, R. 1998. Seeds of ecosystem management in Leopold's A Sand County Almanac. Wildlife Society Bulletin 26:757–760.

Hagen, J. 1988. Organism and environment: Frederic Clements's vision of a unified physiological ecology. In The American Development of Biology, ed. Rainger, R., Benson, K. R., and Maienschein, J., pp. 257–280. Philadelphia: The University of Pennsylvania Press.

Hagen, J. 1992. An Entangled Bank: The Origins of Ecosystem Ecology. Piscataway: Rutgers University Press.

Holling, C. 1973. Resilience and stability of ecological systems. Annual Review of Ecology and Systematics 4:1–23.

Hutchinson, G. E. 1957. Concluding remarks. Cold Springs Harbor Symposium on Quantitative Biology 22:415–427.

Hutchinson, G. E. 1959. Homage to Santa Rosalia, or why are there so many kinds of animals? The American Naturalist 93:145–159.

Interagency Ecosystem Management Task Force. 1995. The Ecosystem Approach: Healthy Ecosystems and Sustainable Economies. Vols. I–III. Washington: White House Council on Environmental Quality.

Jensen, W., and Salisbury, F., eds. 1972. Botany: An Ecological Approach. Belmont: Wadsworth.

Johnston, D. W., and Odum, E. P. 1956. Breeding bird populations in relation to plant succession on the piedmont of Georgia. Ecology 37(1):50–61.

Jørgensen, S., Bastianoni, S., Fath, B., and Muller, F. 2007. A New Ecology: Systems Perspective. Amsterdam: Elsevier Science.

Kapustka, L., and Landis, W. 1998. Ecology: the science versus the myth. Human and Ecological Risk Assessment 4:829–838.

Kiesecker, J., Comendant, T., Grandmason, T., Gray, E., Hall, C., Hilsenbeck, R., Kareiva, P., Lozier, L., Naehu, P., and Rissman, A. 2007. Conservation easements in context: a quantitative analysis of their use by The Nature Conservancy. Frontiers in Ecology and the Environment 5:125–130.

Kricher, J. 2009. The Balance of Nature: Ecology's Enduring Myth. Princeton: Princeton University Press.

Leopold, A. 1921. The wilderness and its place in forest recreational policy. Journal of Forestry 19:718–721.

Leopold, A. 1924. Grass, brush, timber, and fire in southern Arizona. Journal of Forestry 22:1–10.

Leopold, A. 1941. Wilderness as a land laboratory. Living Wilderness 6:3.

Leopold, A., and Schwartz, C. W. 1966. A Sand County Almanac, with Other Essays on Conservation from Round River. New York: Oxford University Press.

Leopold, A. S., Cain, S., Cottam, C., Gabrielson, I., and Kimball, T. 1963. Wildlife management in the national parks. American Forests April:32–63.

Levin, S. 1992. The problem of pattern and scale in ecology: the Robert H. MacArthur award lecture. Ecology 73:1943–1967.

Levin, S., and Paine, R. 1974. Disturbance, patch formation, and community structure. Proceedings of the National Academy of Sciences 71:2744–2747.

Lindeman, R. 1942. The trophic dynamic concept in ecology. Ecology 23:399–418.

Margalef, R. 1963. On certain unifying principles in ecology. American Naturalist 97:357–374.

Marshall, R. 1930. The problem of the wilderness. The Scientific Monthly 30:141–148.

McLaughlin, N. 2005. Conservation easements: a troubled adolescence. Journal of Land, Resources, and Environmental Law 26:47–56.

Meine, C. 1988. Aldo Leopold: His Life and Work. Madison: University of Wisconsin Press.

Miller, C. 2001. Gifford Pinchot and the Making of Modern Environmentalism. Washington: Island Press/Shearwater Books.

Muir, J. 1901. Our National Parks. Boston: Houghton Mifflin.

Nicolson, M. 1990. Henry Allan Gleason and the individualistic hypothesis – the structure of a botanists career. Botanical Review 56:91–161.

O'Neill, R. 2001. Is it time to bury the ecosystem concept? (with full military honors, of course!). Ecology 82:3275–3284.

Odum, E. P. 1953. Fundamentals of Ecology. Philadelphia: W. B. Saunders.

Odum, E. P. 1964. The new ecology. BioScience 14:14–16.

Odum, H. T. 1994. Ecological and General Systems: An Introduction to Systems Ecology. Niwot: University Press of Colorado.

Pickett, S. 1976. Succession: an evolutionary interpretation. American Naturalist 110:107–119.

Pinchot, G. 1900. A Primer of Forestry. Part I: The Forest. Washington: Government Printing Office.

Pinchot, G. 1937. The Training of a Forester. Philadelphia: J.B. Lippincott.

Pinchot, G., and Steen, H. K. 2001. The Conservation Diaries of Gifford Pinchot. Durham: Forest History Society.

Pyne, S. J. 1997. Fire in America: A Cultural History of Wildland and Rural Fire. Seattle: University of Washington Press.

Robbins, W., Ackerman, E., Bates, M., Cain, S., Darling, F., Fogg J., Gill, T., Gillson, J., Hall, E., and Hubbs, C. 1963. A report by the Advisory Committee to the National Park Service on Research. Washington: National Academy of Sciences National Research Council.

Rohde, K. 2005. Nonequilibrium Ecology. Cambridge: Cambridge University Press.

Roper, L. W. 1973. FLO: A Biography of Frederick Law Olmsted. Baltimore: Johns Hopkins University Press.

Sellars, R. W. 1997. Preserving Nature in the National Parks: A History. New Haven: Yale University Press.

Simberloff, D. 1980. A succession of paradigms in ecology: essentialism to materialism and probabilism. Synthese 43:3–39.

Soulé, M. E., and Lease, G. 1995. Reinventing nature? Responses to Postmodern Deconstruction. Washington: Island Press.

Strong, D. H. 1988. Dreamers & Defenders: American Conservationists. Lincoln: University of Nebraska Press.

Tansley, A. 1935. The use and abuse of vegetational concepts and terms. Ecology 16:284–307.

USDI. 1970. Administrative Policies for Natural Areas of the National Park System. Washington: National Park Service, United States Department of the Interior.

Vale, T. R. 2005. The American Wilderness: Reflections on Nature Protection in the United States. Charlottesville: University of Virginia Press.

Warming, E., Balfour, I. B., Groom, P., and Vahl, M. 1909. Oecology of Plants: An Introduction to the Study of Plant-communities. Oxford: Clarendon.

Watt, A. 1947. Pattern and process in the plant community. The Journal of Ecology 35:1–22.

Westman, W., and Peet, R. 1982. Robert H. Whittaker (1920–1980): the man and his work. Vegetatio 48:97–122.

White, P. 1979. Pattern, process, and natural disturbance in vegetation. The Botanical Review 45:229–299.

Wilkins, T. 1995. John Muir: Apostle of Nature. Norman: University of Oklahoma Press.

Wright, H. E., and Frey, D. G., eds. 1965. The Quaternary of the United States: A Review Volume for the VII Congress of the International Association for Quaternary Research. Princeton: Princeton University Press.

Wu, J., and Loucks, O. 1995. From balance of nature to hierarchical patch dynamics: a paradigm shift in ecology. The Quarterly Review of Biology 70:439–66.

Zimmerer, K. 1994. Human geography and the "new ecology": the prospect and promise of integration. Annals of the Association of American Geographers 84:108–125.

Chapter 3
A Thing is Right

The modern concept of the ecosystem is by no means a settled question. Though no one denies that the physical and chemical conditions of an environment change over time or that species migrate in more or less independent ways, there are still those who maintain that the ecosystem is best conceptualized as a holistic unit and protected in its ideal state. Others are equally insistent that there are serious deficiencies with the holistic ecosystem concept and associated management goals. The varieties of ecosystem perspectives in between these extremes make it plain that the possibilities are not a dichotomy but a continuum. In this chapter I present ecosystem views that are representative of the two ends of the conceptual continuum, and then I consider some key tenets of the modern "ecosystem approach" to conservation.

The Adaptive Cycle

Despite the ubiquity of the "progression to climax" diagram, the ecosystem is better represented by the Adaptive Cycle (Gunderson and Holling 2002; Walker and Salt 2006). In this model, an ecosystem exists in cyclic phases (Fig. 3.1). Because they are cyclic, there is no particular beginning or end point, but it is easiest to conceptualize by beginning with the Rapid Growth phase. This may be thought of as the phase of early succession. Imagine an abandoned field, which perhaps was farmed until it was left to revert to a more natural condition. Rather quickly, the field would be invaded by a variety of plant, animal, and microbial species. These early colonizers are known as r-strategists – they are species that disperse well, grow rapidly in a wide range of conditions, and reproduce quickly. These species are not particularly good competitors over the long term, but they thrive in the short term. In some cases, it may be that the pioneer r-strategists have altered the environmental conditions – perhaps they have mobilized soil nutrients, added organic matter to the soil, or shaded the soil surface – and thereby have improved or worsened the colonization conditions for other species.

Species that can tolerate conditions well enough to become established later in the game may be called K-strategists. They disperse, grow and reproduce more slowly, but they are more highly specialized for efficient use of a particular set of resources. This makes K-strategists excellent long term competitors, and given

D. J. Spieles, *Protected Land*, Springer Series on Environmental Management,
DOI 10.1007/978-1-4419-6813-5_3, © Springer Science+Business Media, LLC 2010

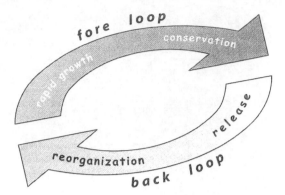

Fig. 3.1 The Adaptive Cycle. From *Resilience Thinking* by Brian Walker and David Salt. Copyright © 2006 Brian Walker and David Salt. Reproduced by permission of Island Press, Washington, DC

enough time they will out-complete the early successional species. The dominance of K-strategists has been called the Conservation Phase of the Adaptive Cycle. It is called the Conservation Phase because of the efficiency of K-strategists, but it is worth noting that this late-successional phase is also the target of many ecosystem conservation efforts. It is seen by some as the stable state of the ecosystem, in which species turnover is low and nutrient cycles are relatively closed. This is the Clementsian climax community, and in this conception it consists not only of organized, specialized, highly adapted K-strategists but also of feedback loops that maintain the system in this state.

The Conservation Phase ends when an ecological disturbance disrupts the conditions that favor the organized community of K-strategists. The disturbance may be abiotic and density independent, like fire, flood, windstorm, or ocean surge; it may be biotic and density dependent, such as the action of herbivores or pathogens. In all of these examples, the disturbance is a sudden and temporary event that alters the biogeochemical conditions of the environment in question. Disturbances vary in scale, frequency and intensity. According to the Adaptive Cycle model, a single, small, low intensity disturbance, like a lone tree falling in a forest, would not be enough to alter the Conservation Phase. A large, intense disturbance, like a major windstorm that uproots a large percentage of the trees in a stand, might change conditions to such a great extent that the K-strategists are no longer favored. The result is a chaotic Release Phase, in which resources and species are highly variable and relatively unpredictable. For example, sunlight, bare soil, decaying organic matter, and mobilized nutrients might suddenly be available after a major windstorm – things that were in short supply under the dense canopy of the mature forest. In the Release Phase there may be an assortment of r- and K-strategists coexisting, including species that arrived via long distance dispersal and those that were present before the disturbance, with specialists alongside generalists. Over time, the Adaptive Cycle hypothesizes that species will begin to assemble themselves anew, in what is called the

Reorganization Phase. The reorganized biota may then progress through the Rapid Growth phase toward a Conservation Phase that is similar to the pre-disturbance Conservation community, or it may assemble into a community with different characteristics. With degrees of variation, the cycle then repeats (Walker and Salt 2006).

The Adaptive Cycle may be interpreted quite rigidly as a predictable cycle that returns deterministically to a particular climax community: the ideal ecosystem. But the concept is also open to stochasticity. One may be a holistic adherent to the Adaptive Cycle and still recognize that the cycle can happen simultaneously but asynchronously in various patches of the ecosystem. The cycle allows for random recruitment of species, novel assemblages, and variation in ecological processes. It is therefore not predicated on the persistence of a climax state; indeed, disturbance is a necessary component of the model. In the context of ecosystem preservation and conservation, though, the Adaptive Cycle can be a goal-oriented concept. By this I mean that a particular phase can be seen as a desirable, thus making the conservation goal the maintenance of that state *in spite* of disturbance.

A number of objections to a rigid, holistic interpretation of the ecosystem in successional cycles have been raised (O'Neill 2001). First, the concept seems to imply a discrete unit with spatial boundaries, when in fact every aspect of the ecosystem has its own boundaries: species have different ranges, functional processes like productivity and decomposition transcend species distribution, geographic boundaries may have little to do with disturbance area and extent, soil and water conditions are often multiple discordant gradients. The objection is that any "phase" of the Adaptive Cycle must be identified by spatial boundaries that are rather arbitrary. Second, it could be argued that any "stable state" is a perception of the observer, with arbitrary temporal boundaries. Beyond the defined time of a particular Conservation Phase, it becomes difficult to identify exactly what characteristics would need to remain unchanged for a post-disturbance recovery to the same conservation state. Essentially the argument is that each iteration of the conservation state – indeed, every state of every phase in the cycle –is unique. It is, therefore, not entirely clear that certain qualities or properties of an ecosystem are *cyclic*. Finally, some find the concept of the desirable stable state to be problematic for ecosystem management in practice. The problem is not with the idea of desirability, for clearly humans have a preference for certain ecosystem properties that might enable commodity harvest, allow for beneficial function, or provide aesthetic pleasure. Rather, the question concerns the management mindset to treat disturbance and release as departures from a state that should be maintained.

As we have seen in the previous chapter, these different perspectives (1) are not new, and (2) form the theoretical basis from which ecosystem management policy has been drawn. As the national ecosystem approach to management indicates, American ecosystem protection has in practice been decidedly holistic despite the objections presented above. Thus the concept of the stable climax state – the ideal Conservation Phase – is alive and well in American ecosystem management.

Diversity, Stability, Health and Integrity

"A thing is right," wrote Aldo Leopold in A Sand County Almanac, "when it tends to preserve the integrity, stability and beauty of the biotic community. It is wrong when it tends otherwise." This oft-quoted line could be considered part of the preamble for the constitution of modern American ecosystem protection. Leopold was a champion for man's harmony with the land – this, in fact, was his definition of conservation. He was an advocate for ecosystem health: "the capacity of the land for self-renewal" (Leopold and Schwartz 1966; Leopold et al. 1999). To advocates of the holistic approach to management, the Leopoldian concepts of ecosystem health and ecological integrity are paramount. They seek, as Leopold did, a stable harmony with the land. At the heart of this view of ecosystem management, in both Leopold's day and ours, is diversity. The diversity of living things in a natural area is considered to be linked with the stability, integrity, and health of the ecosystem. Accordingly, we now consider diversity, stability, health and integrity against the backdrop of the holistic and individualistic debate.

Diversity

The simplest measure of ecological diversity is known as *alpha diversity* (Vane-Wright et al. 1991). Alpha diversity quantifies the number and relative abundance of different types of organisms in a defined location. Though simple, alpha diversity has value; knowledge of alpha diversity over time provides an indicator of ecosystem status. For example, the ecosystem may show a trend toward or away from species homogeneity, or abundance may become more or less evenly spread among constituent species. Alpha diversity may be compared to identify species unique to different regions (this is known as beta diversity) or to quantify the total number of species across a region of different ecosystems (gamma diversity; Whittaker 1972). Ecologists certainly prize alpha diversity; a diverse ecosystem is considered to be of higher quality than a more homogeneous ecosystem. But why? Why is diversity a desirable attribute? There are many compelling reasons, but the reasons are difficult to explain with alpha diversity alone. Simple lists of species are static, sterile, and unable to explain the desirable qualities of the system. In ecology, the effort to understand the implications of organism variety and abundance has necessitated some deeper concepts of diversity.

The ecological concept of *functional diversity* can be traced to early efforts to categorize organisms based not on their appearance but instead on their role in the ecosystem (Elton 1927). Functional classification requires an understanding of each organism's life strategy. For example, an organism's function might include the ways in which it obtains nutrients or its behavior in stressful situations. Functional diversity is not always correlated with alpha diversity, for many different types of organisms may perform similar roles in an ecosystem even as some taxonomically similar organisms have different functions. Functional diversity is a powerful concept, and one that is more difficult to measure or perceive than alpha diversity. To understand functional diversity, one must determine not only the species in the ecosystem but also how those species manage to exist as they do. Invariably, the role of an organism in an ecosystem includes its association with other organisms. This makes functional diversity a measure of connectivity and interdependence in a way that alpha diversity alone is not. Functional diversity is desirable, as the theory goes, because it is associated with a complex range of system properties. Thus, one answer to the of question diversity's importance is that greater functional diversity yields more unique patterns of interaction among organisms, more efficient and complete processing of information and material, and a greater opportunity for symbiotic relationships (Odum 1969).

Ecosystem Services

An anthropocentric reason to value diversity in ecosystems is that functional diversity promotes certain desirable ecosystem functions. The idea that an ecosystem has particular functions hearkens back to the holistic idea of *emergent properties*, that the whole of the ecosystem is greater than the sum of the parts. Undeniably, living systems function in particular ways that are of great value to humans. Ecosystems process and sequester carbon, they retain floodwaters and buffer against storm surge, they cycle minerals, they break down pollutants and decompose wastes. None of these ecosystem services can be attributed to a single species; they require functional diversity in an appropriate abiotic environment. Consider the nitrogen cycle, on which the human race is greatly dependant. Nitrogen fixation – the conversion of atmospheric nitrogen to ammonium – is accomplished by a select few microorganisms which produce enzymes specialized for this function. Ammonium is a useful form of nitrogen for plants, animals and other microbes, but the wastes of these organisms would accumulate to toxic levels were it not for nitrifying bacteria, which convert ammonium to nitrite and nitrate. Likewise, excessive accumulation of nitrate is alleviated by denitrifying bacteria, which convert nitrate back into atmospheric nitrogen. Thus a community of microorganisms provides a valuable service for humankind. Of course, the abiotic setting is just as critical for this cycle as the biota. Nitrification, for instance, requires the presence of oxygen, but denitrification requires the absence of oxygen. Just any setting won't do, and some settings are ideal. The shallow waters of a marsh, in which oxygen is available above the anoxic muck, is one ideal setting, and so a marsh ecosystem is the site of an ecosystem service.

There appear to be species that perform similar, or even the same, functions in ecosystems. Such functional redundancy does not mean that different species are functioning in identical ways in all facets of their existence – this would make them the same species. Instead, the idea is that different species may have some degree of functional niche overlap. For example, an ecosystem may have two different species of nitrogen fixers, one a free-living producer and the other a plant symbiont.

So, while it is true that these species share the function of nitrogen fixation, they really have different functional niches. In a given functional category, such as predator, pollinator, or decomposer, there may be several or many species at any given time performing that function.

In an ecosystem management sense, it is easy to see the advantage of protecting species diversity. More species means a greater variety of function, and that might mean greater functional redundancy. Greater redundancy may be a protection against catastrophic ecosystem failure. Paul and Anne Ehrlich compare this to rivets on the wing of a plane. One may lose a rivet or two and go on flying, but lose enough rivets and the wing falls off (Ehrlich and Ehrlich 1983). Of course, some rivets are located in more critical places on the wing – some species may add relatively little to the sum total of the ecosystem's function, while other species add a great deal more (Peterson et al. 1998). Even so, it is tempting to conclude that more rivets are better than fewer rivets. But this analogy has a limitation. A wing's relationship to an airplane, at least from a passenger's perspective, has but one proper arrangement. What is the proper arrangement of an ecosystem?

The Stable Equilibrium

The ecological mantra that "diversity means stability" has enjoyed support since Darwin, but it has come under scrutiny in recent decades. Early observations by Elton (1958) and MacArthur (1955) supported the logic that ecological communities with fewer species are more susceptible to drastic population fluctuation and invasion by other species. Conversely, species diversity was thought to provide functional redundancy, such that the decline or loss of a particular species is buffered and the overall function of the system remains stable (McCann 2000). But even the definition of stability is unclear. Given the role of change in the Adaptive Cycle, what does it mean to have a stable ecosystem? That its overall biomass remains unchanged over time? That the dominant species remain dominant? That the density of each population in the ecosystem – or of the overall system – remains unchanged? Or do we mean that the rates of ecological processes, like photosynthesis, productivity, decomposition, and nutrient assimilation remain unchanged?

Myriad definitions of ecosystem stability exist – there are, by one count, "163 definitions of 70 different stability concepts" (Grimm and Wissel 1997). Stability can be taken to mean constancy of populations and processes within defined bounds; it can mean that some measure of the whole ecosystem, like aggregate species density, gravitates toward one or more values; it can be thought of as resistance to change and persistence in the face of disturbance; or it can be synonymous with resilience, meaning a return to a defined state after perturbation. Generally these definitions of stability are all based on the idea of an ecological state of *equilibrium*. The ecological equilibrium is "a particular system state at which all the factors or processes leading to change are being resisted or balanced" (Wu and Loucks 1995). Many scholars have noted that the stable equilibrium concept has historical roots in the "balance of nature" concept of previous centuries. As we have seen, the cybernetic analogy of nature, based on nineteenth and twentieth century physical sciences, and the homeostatic analogy, drawn from physiology, were two inspirations for ecosystem ecology. Both conceptualize the ecosystem as a self-regulating unit, and both postulate that the state at which the system is regulated is a stable equilibrium.

Aside from its ambiguity, the problem with the stable equilibrium is that it does not exist, at least not for long. It is clear that static equilibrium – complete constancy of populations and interconnections over time – does not occur in ecological communities. *Dynamic* equilibrium is the idea that the equilibrium state is maintained as opposing variables change at the same rate to oppose and balance one another. A classic example is that of the predator–prey oscillation: when prey numbers increase, predator numbers increase accordingly and thus maintain the balance of the respective populations. The problem comes in the definition of which ecosystem properties are to be considered and the boundaries in which their variation is deemed stable.

Indeed, an ecosystem may have a stable state, or even multiple stable states. A mangrove swamp that has existed for as long as humans have been noticing mangrove swamps, upon being devastated by a hurricane, may re-assemble into a mangrove swamp. A beech-maple forest in the Midwestern United States, subjected to a devastating windstorm, may still appear to be a textbook example of a beech-maple forest. But are the Conservation Phases with which we identify these ecosystems truly "stable equilibria" or are they human perceptions of stability? Consider the spatial scale of the beech-maple forest. On the scale that we perceive the forest – perhaps hundreds of acres – it appears that the forest recovers from the windstorm and that the dominant species have regained their dominance. How about on a much smaller scale? The characteristics of the soil fauna may be dramatically different post-windstorm – the soil is receiving more sunlight, it is drier, there is suddenly abundant coarse woody debris. Meanwhile, fugitive species have arrived on the scene – plants, animals, and microbes – and rapidly extract nutrients, excrete wastes, reproduce and die. Another nearby patch within the same forest may have different characteristics. The ecosystem, perceived at this scale, has been greatly altered by the storm, even after the beech and maple trees have returned. What if we consider the temporal scale? Stability, as it is usually perceived, fits neatly into the time scale of a human lifetime. But several centuries ago, the ranges of beech and maple trees did not coincide as they do now. What can it mean, then, to say that the beech-maple community is at a stable equilibrium? Aren't "stable states" entirely dependent on the human definition?

An individualist might argue that ecosystem properties and processes vary at multiple scales in response to environmental fluctuation, and since environmental conditions are rarely stable, ecosystems are rarely stable. Human attempts to draw boundaries around variation have resulted in such innovations as the "loose equilibrium" and "clouds" of stable states (Collins 2000; Belovsky 2002); these are, in the words of Robert O'Neill (2001), "putting splints and patches on an old horse."

With conveniently defined boundaries and scale, we can identify stable equilibria of any ecosystem. But given the role of human perception, should it be an objective of our approach to managing natural areas – the ecosystem approach – to achieve and maintain stability? Stability is a human construct, like the ecosystem itself, and as such it occurs only in the eye of the beholder. Most definitions of stability seem to suggest that stability is the maintenance of an ecosystem in some optimal state. Even if we accept these definitions, the relationship of ecosystem stability to the diversity of living things in the ecosystem – long considered an important management parameter – is not clear. If we insist that ecosystem stability is our goal, does this then mean that maximizing diversity will always achieve that end? Mounting evidence suggests that it does not. Species-rich systems may be susceptible to change, and species poor systems may be remarkably constant in certain attributes of structure and function. Species diversity alone is seldom the sole determinant of system stability at any level (Golley 1996; Ives and Carpenter 2007).

And so, if stability is subjective and transitional and its relationship with diversity is so nebulous, is a reasonable expectation for a patchy ecosystem in a fluctuating environment?

Ecosystem Health

Along with stability, a second objective of the ecosystem approach is ecological health – Leopold's "capacity for self-renewal." A healthy ecosystem "produces biomass; recruits, maintains, and cycles nutrients; holds the soil; modulates the flow of water; and maintains other ecosystem processes" (Callicott 2000). The term *health* evokes an image of the individual organism, and indeed the concept of ecosystem health has a long historic connection with the idea of the ecosystem as a superorganism. The modern conception of ecosystem health does not require that the ecosystem be considered a superorganism, but it does presuppose that the ecosystem is a definable entity and that it has characteristic functions and processes. Ecosystem health, then, is all about determining whether or not the processes and functions are what they should be. Peter Callow (1992) notes that the health analogy, it its "weak" form, identifies the "normal" conditions of an ecosystem; whereas the "strong" form of the analogy "defines a condition favorable (i.e. optimal) for the functioning of the whole organism." And what are the "normal" conditions of an ecosystem? We may define "normal" by observation of similar ecosystems that are relatively free from stress – and thus "healthy," in a bit of circular logic – or, if data are available, to the ecosystem in question itself, from a period before degradation. In the strong form of the analogy, the "optimal" state may well be defined as the state that is most beneficial, aesthetically pleasing, or convenient to humans.

Whether we call the healthy state "normal" or "optimal," it is clear that ecosystem health is linked with the concept of stability. A healthy ecosystem, one might argue, has processes and functions that exist in a dynamic equilibrium; the healthy ecosystem persists in the equilibrium state and returns to the equilibrium state after perturbation. In the strong holistic view of ecosystem health, there is emphasis on the persistence of the system in an optimal state: in James Karr's (Karr et al.1986) language "a biological system … can be considered healthy when its inherent potential is realized, its condition is stable, its capacity for self-repair when perturbed is preserved, and minimal external support for management is needed." Thus, to determine whether or not an ecosystem is healthy, one needs to define this "inherent potential," this "normal" stable state.

Some have argued that there is no scientific basis for defining the healthy state of an ecosystem. Consider the assessment of Robert Lackey (2001), for instance: "At the core of the debate over ecosystem health are a number of implicit, but highly contested, value-based assumptions that masquerade as science." Often, naturalness, i.e. the state of being unaltered by humans, is perceived as "normal," and thus the healthy state – but of course the "natural state" is also a perception of the human observer. The natural or normal state, for example, may be defined as the condition of being within range of historic processes and functions for that particular ecosystem (Callicott 1995). But as we have seen, ecosystems have undergone wild changes structurally and functionally since the last glaciation, such that they can hardly be referred to as units with a single natural state. The historical state of normalcy must be defined arbitrarily. In North America, it is often the condition that was first observed by men with white skin.

If a particular ecosystem had a single stable state, it might be deemed unhealthy if its key processes, like nutrient cycling, productivity, and symbioses, deviated significantly from that stable state. If the ecosystem had multiple stable states, one state might be deemed healthy, and the others unhealthy. For example, a coral reef might be considered healthy if the coral is actively growing, if the diversity of marine organisms in and around the reef is high, and if there are complex symbiotic pathways through which nutrients flow. As we shall see in Chap. 10, excessive stress has been known to trigger a shift to a different state of the reef ecosystem, e.g. one dominated by algae. The diversity and interconnectedness of species decline, the coral dies, and foods webs are simplified. The ecosystem has changed, and a vacationing scuba diver may quickly determine that this is an unhealthy reef, as might the fishing industry that depends upon the reef as a breeding ground.

The label "unhealthy" is a matter of human perception. There is nothing inherently wrong with the new state of the ecosystem. In addition, there is nothing particularly stable about either state, other than the presence of dominant species within a human-defined spatial and temporal scale. True, in a policy-making sense, there may be some merit to defining the healthy state of the reef or any ecosystem; it is the state that human society desires, and it may serve as a reference point for gauging the effects of anthropogenic stressors like over-harvest and eutrophication. It may also provide an objective for management efforts that seek to maintain the ecosystem in a state that is useful or attractive to humans. But there is no scientific basis for considering one defined state healthy and another unhealthy, and there may actually be harm in working to maintain an ecosystem in a particular state.

In spite of this logic, management that is based on ecosystem health typically targets the ideal state as its sole objective. One early effort to define the criteria by which we may measure ecosystem health lists several "critical ecosystem characteristics." Among them are: "habitat for desired diversity, a robust food chain supporting the desired biota, an adequate nutrient pool for desired organisms, adequate nutrient cycling to perpetuate the ecosystem, adequate energy flux for maintaining the trophic structure, feedback mechanisms for dampening undesirable oscillations" (Schaeffer et al. 1988). A decade later, David Rapport and others softened the "ideal state" language somewhat and reclassified the characteristics of ecosystem health into "vigor, organization, and resilience" (Rapport et al. 1998). Thus, to paraphrase: a highly productive, metabolically active community with diverse interconnections that is able to maintain its structure and function in the presence of stress is considered a healthy ecosystem. One may ostensibly assess ecosystem health under this definition by measuring primary productivity or rates of assimilation or decomposition (greater is healthier), trophic guild diversity (more is healthier), and the constancy of structural pattern and functional processes (more stable is better). In unabashed holism, the healthy ecosystem is "stable and sustainable, maintaining its organization and autonomy over time" (Costanza et al. 1992).

Ecological Integrity

In 1995, Dan Wicklum and Ronald Davies passed harsh judgment on the concepts of ecosystem health and integrity: "The phrases ecosystem health and ecosystem integrity are not simply subtle semantic variations on the accepted connotations of the words health and integrity. Health and integrity are not inherent properties of ecosystems." And yet, the ecosystem approach and much federal environmental legislation specifically mandates the restoration and maintenance of the physical, chemical, and biological integrity of ecosystems. What is meant by ecosystem integrity? Ecological (or biological) integrity, like stability and health, does not have a single, clear definition. According to an early definition by David Frey (1977), integrity is "the capability of supporting and maintaining a balanced, integrated, adaptive community of organisms having a composition and diversity comparable to that of the natural habitats of the region." Callicott (1995) suggests that integrity should be a community-oriented concept: "preserving native species populations, in their characteristic numbers, with their evolved or historic interactions." In both of these definitions, it is clear that integrity concerns the composition of the community in the ecosystem, as opposed to health, which is more about function and process. But community composition and system function are not mutually exclusive. Thus, the US EPA (US EPA 1990) defines biological integrity for aquatic ecosystems as "the condition of the aquatic community that inhabits unimpaired water bodies of a specified habitat as measured by community structure and function." The Nature Conservancy suggests that an ecosystem has integrity "when its dominant ecological characteristics (e.g. elements of composition, structure, function, and ecological processes) occur within their natural ranges of variation and can withstand and recover from most perturbations imposed by natural environmental dynamics or human disruptions" (Parrish et al. 2003). Clearly, the EPA and TNC definitions meld biological integrity to ecosystem health by combining community composition and ecological function. To Orie Loucks (2000), integrity can be measured entirely as function, and not as composition at all, by comparing an ecosystem's basic emergent processes (like productivity, decomposition, and nutrient cycling) with reference ecosystems. James Karr (2000) sees biological integrity as one end of a continuum of ecological health; it is the pristine condition, the state that has not been degraded by human influence. In this sense, integrity is the epitome of health; "a benchmark of biological condition."

The wide selection of definitions has not prevented ecologists from quantifying ecological integrity. Some of the early indices of biological integrity were designed for aquatic ecosystems. In 1981, Karr presented the IBI (Index of Biotic Integrity) for assessing the integrity of fish communities. The index is based upon compositional variables, including species richness and composition. In ensuing years, the number and variety of integrity indices has exploded, including indices for communities of aquatic insects, birds, amphibians, plants, diatoms, reptiles and so on (Table 3.1). Naturally, each index must be specific to region (one cannot use an integrity index in Alaska if it was developed for Florida) and to a particular type of ecosystem (e.g. small warm water stream, central Appalachian highlands, freshwater emergent marsh). Each index must also have one or more reference ecosystems as a model of biological integrity. Generally, the reference system is high in alpha diversity, relatively free of anthropogenic stress (like deforestation, pollution, or fragmentation) and in particular supports a diversity of species that are considered native to that region.

When ecosystem integrity is based on the plant community, the presence of native species is often an important attribute. The goal of achieving "native species in their characteristic numbers"… "comparable to that of natural habitats" is of course more confounded than it first seems. By presidential executive order in 1999, a native, or indigenous, species for a particular ecosystem was defined as "a species that, other than as a result of an introduction, historically occurred, or currently occurs in that ecosystem." The Forest Service definition links "native species" with their evolution: Native species are "all indigenous, terrestrial, and aquatic plant species that evolved naturally in an ecosystem" (Federal Register Vol. 73 No. 30 2008). I belabor these definitions to make the point that they are written from the perspective of the stable climax community. If a species was here historically, the reasoning goes, then it must belong here; if it is part of this community now and was not brought here by modern humans, it must have evolved with this community. As we have seen, of course, species can and do migrate independently and continuously, so "historically occurred" is somewhat imprecise; in North America, it is often understood as the point of European American arrival. Charles Mann (2005) has made a strong case that the "nature" encountered by European American settlers was hardly pristine or stable. Pinpointing a particular location as the home of evolution for a particular species seems even less meaningful. Species have evolved over enormous spatial and temporal time scales, and there is no evidence to suggest that community assemblages remain constant over evolutionary time. Further, as Mark Sagoff (2005) has pointed out, the evidence that native species are the only suitable members of an ecosystem is not apparent – nor has it been clearly demonstrated that nonnative species "behave in general any differently than native ones."

Table 3.1 Characteristics of representative indices of biological integrity

Integrity index	Total species	Community heterogeneity	Native/nonnative species	Specialized organisms	Stress tolerant organisms	Similarity to reference	Indicator Organism	Guild
Index of Biotic Integrity (IBI; fish)	X				X		X	X
Invertebrate Community Index (ICI)	X				X		X	
Floristic Quality Assessment Index (FQAI)	X		X	X				
Index of Plant Community Integrity (IPCI)	X	X	X	X				
Amphibian	X				X			
Trophic Diatom Index (TDI)		X			X		X	
RDI (reptiles)	X	X				X		X
Bird Community Index (BCI)	X			X				X

Sources: IBI and ICI, Karr 1991; FQAI, Cronk and Fennessy 2001; IPCI, DeKeyser et al. 2003; Amphibian, Micacchion 2002; TDI, Kelly and Whitton 1995; RDI, Thompson et al. 2008; O'Connell et al. 1998

Despite these conceptual conundrums, species that are defined as native typically increase an ecosystem's biological integrity score, while the presence of a nonnative species is taken as an indication of low integrity. To supplement or supplant nativity, most indices of biological integrity include a broader assessment of species quality. For example, species may be assigned a rating based on its specificity – its fidelity to a particular ecological condition. Specialist species are awarded more points than generalist species, and particular species or guilds are prized. Integrity indices that are based on the presence of specialists, degree of similarity to a reference system, or the presence of indicator organisms or guilds are clearly equating integrity with an ideal late Conservation Phase community. There are also more individualistic indicators of integrity, such as stress tolerance and community heterogeneity, suggesting that individualism and integrity might not be incompatible.

If ecological integrity and health are so ingrained in the American concept of ecosystem protection, why would Wicklum and Davies (1995) question their use? Very simply, they see both concepts as scientifically unsound. To paraphrase: Ecosystem health is an invalid analogy based on human health. It defines an optimum condition from which any change or deviation is by definition negative. Such an optimal state may easily be identified for organisms, but ecosystems are not organisms. Ecological integrity, according to these authors, is equally inappropriate. It is predicated on a predefined baseline state that is static and subjective and on the concept that the baseline state can organize and maintain itself through self-correcting processes. They conclude, quite simply, that "no such processes and no such state or conditions exist for ecosystems."

Stability, Health and Integrity in Perspective

Calow's "weak" and "strong" modifiers for the concept of ecosystem health can also be applied to ecosystem stability and integrity, just as I applied them to holism in the Chap. 2. In the strongest sense, health, stability and integrity are referring to the embodied attributes of an ideal ecosystem – functioning in an optimal way, maintaining that function at a stable equilibrium, populated with native, late-successional species in characteristic abundance. This is not far removed from the Clementsian climax. Given the evidence that species' association and organization are coincidental and temporary, this "strong" view of ecosystem health, stability, and integrity might be rejected on the grounds that these concepts are all human constructs.

I suspect that few scholars remain at the extreme view of the Clementsian superorganism. Modern proponents of the climax community admit that there is no such thing as perfect constancy, balance, or equilibrium in nature, just as individualists might acknowledge that "structure, pattern, and predictability" may be found in nature (Holland 2000; Partridge 2000). But even in a "weak" form, health, stability, and integrity imply that there is a proper condition or set of characteristics for an ecosystem – the normal condition, as determined by comparison with a normal reference ecosystem. This is still unacceptable, the individualist might argue, for the definition of normal is arbitrary. Granting this point, advocates of ecosystem health, stability and integrity might be willing to go still weaker, so that normal refers to the ecosystem's "undiminished ability to continue its natural path of evolution, its normal transition over time, and its successional recovery from perturbations" (Pimentel et al. 2000) and stability refers to a "pattern of fluctuations" in a defined space and time (Partridge 2000). If, in this weakened form, the holistic view is all about evolution, transition, perturbation, and fluctuation, there may not be much of an argument here after all.

My purpose is neither to heal old wounds nor to open new ones, but rather to consider where American ecosystem protection and management fall on the "strong" to "weak" continuum (Fig. 3.2). As we have seen, the ecosystem approach of our land protection agencies advocates ecosystem health, integrity, and management of the ecosystem toward an ideal state even while acknowledging that ecosystems are constantly changing. In practice, our most intensively managed ecological preserves – the Oak Openings savanna or the Flint Hills prairies, for example – tend to be managed as though there is a stable state that must be preserved and maintained. Clearly, there is a strong holistic idealism in this type of management. In some cases, it seems that we are trying to preserve the ecosystem in a Conservation Phase that is equivalent to the climax forest. In other cases, like grassland preservation, we try to prevent succession to a particular Conservation Phase because an intermediate successional phase is more desirable. Despite the rise of nonequilibrium ecology, management that targets a particular ecosystem arrangement for preservation is persistent.

At the heart of this mindset are the concepts of ecosystem health, integrity, and stability. As a case in point, consider a bit of new federal legislation called "America's Wildlife Heritage Act," proposed in 2009. This bill would require federal land managers to "maintain sustainable populations of native species and desired non-native species," as indicated by a set of focal species that "provide insights to the integrity of the ecological systems to which they belong." It is as though the Leopoldian "thing that is right" has been identified as the state of stable health and

**The Ecosystem
Conceptual Continuum**

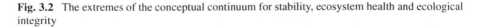

Radically Individualistic

- Multiple, transient, ephemeral equilibria
- Ecosystem function in stochastic fluctuation
- Community consists of tolerant species with chance opportunity

Strongly Holistic

- Stable equilibrium as conservation goal
- Identifiable state of optimal function (ecosystem health)
- Community consists of normal species in characteristic abundance

Fig. 3.2 The extremes of the conceptual continuum for stability, ecosystem health and ecological integrity

integrity – with proper organisms in optimal arrangement – and "things that tend otherwise," like individualistic, stochastic succession, are wrong.

And what is the problem with striving to maintain ecosystems in an optimal, ideal state? To answer this, let me return to the importance of biological diversity, which might be a point of common agreement for the holist and the individualist. Given the value that ecologists and conservationists place on biological diversity, it would seem logical to protect the drivers of diversity. What drives diversity? This, too, is debatable, though there is evidence for several factors (Krebs 2009). Some of these factors might be beyond the control of an ecosystem manager: a larger the geographic area, for example, is likely to support a greater diversity of species. Similarly, warm temperatures and ample water and solar energy are associated with greater species diversity. In a given area, there are some factors of diversity that can be manipulated by humans; among these are the disturbance regime, the heterogeneity of habitat, and the degree of interspecific interaction. Now consider our most holistically managed ecological preserves. The management goals of these protected ecosystems commonly include maintenance of the ideal state by encouraging desirable species, both by manipulating the abiotic (e.g. creating a regular disturbance regime that is ideal for the desirable species) and the biotic (e.g. excluding undesirable competitors). I suggest that in this well-meaning effort, we are diminishing three important drivers of diversity. First, by using disturbance as a maintenance tool, we often regularize the disturbance regime, and hence select for a narrow ecological demographic. Second, by striving to maintain the ecosystem in a particular phase of the Adaptive Cycle we minimize the patchy mosaic of successional stages and thereby limit habitat heterogeneity. Third, by managing exclusively for a desirable set of species we limit competition – another potential driver of diversity.

It is true that we can, through management, maintain an ecosystem in something of an ideal state. In so doing, we may achieve relative stability, we may maintain functions that constitute ecosystem health, and we may protect the specialized native species that signify ecological integrity. But the conditions that make our preserved ecosystem possible are impermanent, and the capacity of the protected area to respond to change is a function of its biological diversity. If, in our efforts to preserve an ideal ecosystem state, we have limited the ecosystem's capacity for response, what will be the consequences when conditions change?

References

Belovsky, G. 2002. Ecological stability: reality, misconceptions, and implications for risk assessment. Human and Ecological Risk Assessment 8:99–108.

Callicott, J. 1995. A review of some problems with the concept of ecosystem health. Ecosystem Health 1:101–112.

Callicott, J. 2000. Harmony between men and land: Aldo Leopold and the foundations of ecosystem management. Journal of Forestry 98:4–13.

Callow, P. 1992. Can ecosystems be healthy? Critical considerations of concepts. Journal of Aquatic Ecosystem Health 1:1–5.

Collins, S. 2000. Disturbance frequency and community stability in native tallgrass prairie. The American Naturalist 155:311–325.

Costanza, R., Norton, B. G., and Haskell, B. D. 1992. Ecosystem Health: New Goals for Environmental Management. Washington: Island Press.

Cronk, J. K., and Fennessy, M. S. 2001. Wetland Plants: Biology and Ecology. Boca Raton: Lewis Publishers.

DeKeyser, E., Kirby, D., and Ell, M. 2003. An index of plant community integrity: development of the methodology for assessing prairie wetland plant communities. Ecological Indicators 3:119–133.

Elton, C. 1927. Animal Ecology. London: Sidgwick & Jackson, LTD.

Elton, C. 1958. The Ecology of Invasions by Plants and Animals. London: Methuen.

Ehrlich, P., and Ehrlich, A. 1983. Extinction: the causes and consequences of the disappearance of species. New York: Random House.

Frey, D. 1977. Biological integrity of water: an historical approach. In The Integrity of Water. Proceedings of a Symposium, March 10–12, 1975, ed. Ballentine, R. K., and Guarraia, L. J., pp. 127–140. Washington: United States Environmental Protection Agency.

Holland, A. 2000. Ecological integrity and the Darwinian paradigm. In Ecological Integrity: Integrating Environment, Conservation, and Health, ed. Pimentel, D., Westra, L., and Noss, R. F., pp. 45–60. Washington: Island Press.

Ives, A., and Carpenter, S. 2007. Stability and diversity of ecosystems. Science 317:58–62.

Grimm, V., and Wissel, C. 1997. Babel, or the ecological stability discussions: an inventory and analysis of terminology and a guide for avoiding confusion. Oecologia 109:323–334.

Golley, F. 1996. A History of the Ecosystem Concept in Ecology: More than the Sum of the Parts. New Haven: Yale University Press.

Gunderson, L., and Holling, C. 2002. Panarchy synopsis: understanding transformations in human and natural systems. Washington: Island Press.

Karr, J. 1981. Assessment of biotic integrity using fish communities. Fisheries 6:21–27.

Karr, J. 1991. Biological integrity: a long-neglected aspect of water resource management. Ecological Applications 1:66–84.

Karr, J. 2000. Health, integrity, and biological assessment: the importance of measuring whole things. In Ecological Integrity: Integrating Environment, Conservation, and Health, ed. Pimentel, D., Westra, L., and Noss, R. F., pp. 209–226. Washington: Island Press.

Karr, J., Fausch, K. D., Angermeier, P. L., Yant, P. R., and Schlosser, I. J. 1986. Assessing biological integrity in running waters: a method and its rationale. Special publication No. 5. Champaign: Illinois Natural History Survey.

Kelly, M., and Whitton, B. 1995. The trophic diatom index: a new index for monitoring eutrophication in rivers. Journal of Applied Phycology 7:433–444.

Krebs, C. J. 2009. Ecology: The Experimental Analysis of Distribution and Abundance. San Francisco: Pearson Benjamin Cummings.

Lackey, R. 2001. Values, policy, and ecosystem health. BioScience 51:437–443.

Leopold, A., and Schwartz, C. W. 1966. A Sand County Almanac, with Other Essays on Conservation from Round River. New York: Oxford University Press.

Leopold, A., Callicott, J. B., and Freyfogle, E. T. 1999. Aldo Leopold: For the Health of the Land: Previously Unpublished Essays and Other Writings. Washington: Island Press/ Shearwater Books.

Loucks, O. L. 2000. Pattern of forest integrity in the eastern United States and Canada: measuring loss and recovery. In Ecological Integrity: Integrating Environment, Conservation, and Health, ed. Pimentel, D., Westra, L., and Noss, R. F., pp. 177–190. Washington: Island Press.

MacArthur, R. 1955. Fluctuations of animal populations and a measure of community stability. Ecology 36:533–536.

Mann, C. C. 2005. 1491: New Revelations of the Americas Before Columbus. New York: Alfred A. Knopf.

McCann, K. 2000. The diversity-stability debate. Nature 405:228–233.

Micacchion, M. 2004. Integrated Wetland Assessment Program. Part 7: Amphibian Index of Biotic Integrity (AmphIBI) for Ohio Wetlands. Ohio EPA Technical Report WET/2004-7.

Columbus: Ohio Environmental Protection Agency, Division of Surface Water, Wetland Ecology Group.

O'Connell, T., Jackson, L., and Brooks, R. 1998. A bird community index of biotic integrity for the mid-Atlantic highlands. Environmental Monitoring and Assessment 51:145–156.

Odum, E. P. 1969. The strategy of ecosystem development. Science 164:262–270.

O'Neill, R. 2001. Is it time to bury the ecosystem concept? (with full military honors, of course!). Ecology 82:3275–3284.

Parrish, J., Braun, D., and Unnasch, R. 2003. Are we conserving what we say we are? Measuring ecological integrity within protected areas. BioScience 53:851–860.

Partridge, E. 2000. Reconstructing ecology. In Ecological Integrity: Integrating Environment, Conservation, and Health, ed. Pimentel, D., Westra, L., and Noss, R. F., pp. 79–98. Washington: Island Press.

Peterson, G., Allen, C., and Holling, C. 1998. Ecological resilience, biodiversity, and scale. Ecosystems 1:6–18.

Pimentel, D., Westra, L., and Noss, R. F. 2000. Ecological Integrity: Integrating Environment, Conservation, and Health. Washington: Island Press.

Rapport, D., Costanza, R., and McMichael, A. 1998. Assessing ecosystem health. Trends in Ecology & Evolution, 13:397–402.

Sagoff, M. 2005. Do non-native species threaten the natural environment? Journal of Agricultural and Environmental Ethics 18:215–236.

Schaeffer, D., Herricks, E., and Kerster, H. 1988. Ecosystem health: I. Measuring ecosystem health. Environmental Management 12:445–455.

Thompson, S., Thompson, G., and Withers, P. 2008. Rehabilitation index for evaluating restoration of terrestrial ecosystems using the reptile assemblage as the bio-indicator. Ecological Indicators 8:530–549.

Walker, B.H., and Salt, D. 2006. Resilience Thinking: Sustaining Ecosystems and People in a Changing World. Washington: Island Press.

Whittaker, R. H. 1972. Evolution and measurement of species diversity. Taxon 21:213–251.

Wicklum, D., and Davies, R. 1995. Ecosystem health and integrity? Canadian Journal of Botany 73:997–1000.

Wu, J., and Loucks, O. 1995. From balance of nature to hierarchical patch dynamics: a paradigm shift in ecology. The Quarterly Review of Biology 70:439–66.

US EPA. 1990. Biological Criteria: National Program Guidance for Surface Waters (EPA-440/5-90-004). Washington: United States Environmental Protection Agency.

Vane-Wright, R., Humphries, C., and Williams, P. 1991. What to protect? Systematics and the agony of choice. Biological Conservation 55:235–254.

Chapter 4
Disturbance, Stress and Resilience

In *Wonderful Life* (1989), Stephen Jay Gould played a metaphorical game he called "replaying life's tape." In the game, you rewind the history of life on Earth to some point in the past, and then "let the tape run again and see if the repetition looks at all like the original." He was writing about the evolution of various forms of life on Earth, not about the assembly of ecological communities, but the same game may be played with ecosystems. Rewind the tape a short time, perhaps a few million years, and press play. When the tape has returned to the present – assuming humans are still present – would we still be trying to preserve the oak savannas south of Lake Erie? Would we be managing ecosystems in eastern Kansas to favor big bluestem and Indian grass and to discourage woody species? And (though I would like to believe that the channelization of the Kissimmee would not happen in any other reality), would we still by trying to restore a mean annual dry season density of long-legged wading birds on the Kissimmee floodplain to greater than 30.6 birds/km²?

It is, as Gould acknowledges, an experiment that can never be run. But his point is still well made: much of evolution is contingent upon chance events. "Alter any early event, ever so slightly and without apparent importance at the time, and evolution cascades into a radically different channel" (Gould 1989). The same could be said about ecosystems, these coincident assemblages of species and abiotic conditions. Both the assemblage and the conditions are as we perceive them to be because of a history of contingencies. Given the randomness of natural events, it would be difficult to imagine the same ecosystems in the same places after a replay of history. Of course, this does not mean that that the ecosystems we seek to protect are not precious, and neither does it mean that there is no predictability to natural systems (Holland 2000). What it does mean is that the ecosystems we see today – like the species we see today – appear as they do in large part due to random events of destruction, selection, and opportunity. In short, ecosystems are the products of a long and complex history of disturbance and response. Accordingly, it makes little sense to manage our ecosystems as though we must maintain the ideal state that existed when we first encountered them. Rather, we should manage their capacity for response to the next disturbance, whatever that might be.

The idea that disturbance should be part of ecosystem preservation and management is not new – it was, after all, urged by the Leopold and Robbins reports of the 1960s (Chap. 2) – but the role of disturbance and response depends upon one's perspective. In the nonequilibrium view disturbance is not an impediment to ecosystem maturity but rather a driver of biological diversity and spatiotemporal dynamics. Comparatively, the holistic, climax-oriented view of ecosystems generally accepts that disturbance is an important part of the adaptive cycle. Indeed, as an internal biological feedback mechanism the disturbance may be seen as a restorative event. But external disturbances, in the strict holistic view, are something to be resisted, something that the system must recover from. As I have shown, the holistic and individualistic views are not necessarily mutually exclusive, but they fundamentally differ on the use of disturbance in management when one considers how, why, and to what end.

Disturbance

Before discussing the role of disturbance in ecosystem management it might be useful to begin with a formal definition. According to Pickett and White (1985), "a disturbance is any relatively discrete event in time that disrupts ecosystem, community, or population structure and changes resources, substrate availability, or the physical environment." As these authors note, in this definition there is no indication of departure from a "normal" ecosystem state. It does imply, however, that there was "structure" in place prior to the disruption, and it also implies that the change wrought by the event is somehow different than the change that is always occurring in the environment. The holistic interpretation of this definition simply accepts that there is a "nominal state" from which departure has occurred, and even claims that it makes no sense to speak of a disrupted system unless one recognizes the nominal system (Odum et al. 1979). Whether a normal state is recognized or not, the basic definition makes it clear that an ecological disturbance is a disruptive event. But this simple definition lacks reference to scale, and scale would appear to be necessary. For example, a dead branch falling to the forest floor may fit the definition of disturbance for the soil microbe community on which it comes to rest, but it would be hard to argue that the falling branch is a disturbance at the scale of the entire forest.

Pickett and others have attempted to clarify the definition using hierarchies of scale. In their concept, an ecological entity that might be subject to disturbance must have a structural organization that is to some degree persistent. An ecological disturbance, then, is "a change in the minimal structure [of ecological entities] caused by a factor external to the level of interest" (Pickett et al. 1989). In our falling branch example, the level of interest might be the microbial community on a sunlit, warm, dry patch of forest floor. Over a certain time scale, the microbial community might have been relatively persistent – there were dominant organisms in a particular structural arrangement processing nutrients in rates that fluctuated about a mean. One might even use the word *stable* to describe this situation. Now comes the branch, and suddenly, on that particular patch of ground, the sunlight has been obscured. The soil begins to cool and moisten. The form of carbon in greatest abundance has changed, as have the organismal characteristics best suited for the environment. At this scale, a disturbance has occurred. This modified definition acknowledges that there was indeed structure and function in place prior to the disturbance event, and it accounts for scale. It has the added advantage of allowing for relative persistence and organization without requiring a "normal" state from which to deviate or return. Rather than a departure from an ideal state, then, one may see a disturbance event as a force of selection. It is destructive, to be sure – "a discrete, punctuated killing, displacement, or damaging" – and therefore likely to be detrimental for some organisms or species. But it also "directly or indirectly creates an opportunity for new individuals (or colonies) to become established" (Sousa 1984).

Disturbance and Stress

Disturbance has been conflated with the term *stress* often enough that the two terms deserve attention here. Odum (1985) defined ecological stress as "a detrimental or disorganizing influence" in an ecosystem and differentiated stress from a perturbation (aka disturbance), which is potentially beneficial. Odum envisioned stress and disturbance as different levels on the same gradient. At a particular level of disturbance, in Odum's estimation, processes and states of the normal ecosystem may be enhanced. In these cases, the disturbance has acted as a subsidy for the system, improving its performance and ostensibly driving it toward some optimal state. On the other hand, a disruptive influence that is detrimental to the state and processes of the normal system is a stress. To Odum, stress forces the ecosystem from its mature state toward simpler developmental stages – it causes a "breakdown in homeostasis." Odum's "subsidy-stress gradient" (Odum et al. 1979) is dependent on the homeostatic analogy and on subjective definitions of "normal" and "optimal;" as such, his definition of stress may be a difficult pill for the individualist to swallow.

Odum's concept of disturbance and stress has clear ties to the organismal view of the ecosystem – stress is something that strains homeostasis and diverts the system from its optimal steady state climax (Barrett and Rosenberg 1981). For those uncomfortable with this "strong" ecosystem concept, stress and disturbance are less about the maintenance of an optimal state and more about the capacity for organismal response to change. This has been nicely framed by Grime (2001), who differentiated stress from disturbance in plant communities, classifying stress as "the external constraints which limit the rate of dry matter production of all or part of the vegetation." Disturbance, in Grime's scheme, is "the mechanisms which limit the plant biomass by causing its partial or total destruction." Most definitions of stress characterize it as different than disturbance in that (1) stress is more of a chronic condition than a disturbance event, which is relatively temporary and discrete – though the effects of disturbance event may be long lasting; (2) stress impairs the physiological function of organisms on a continuing basis, whereas a disturbance reorganizes ecosystem structure and then allows function, albeit altered function, to proceed; and (3) disturbance creates opportunities for response, while stress limits the range of response (Barrett and Rosenberg 1981).

Ecosystem stressors may include physical restructuring of an environment, as in drainage or erosion, over-harvesting, accumulation of toxic substances, waste, or nutrients, or the introduction of exotic species (Table 4.1). Stress may also refer to an existing environmental condition that limits growth or productivity; thus nutrient limitation, soil or water pH, desiccation, osmotic pressure, and hypoxia may all be considered ecosystem stressors (Rapport and Whitford 1999; Walker 1999). It is not surprising that the concepts of disturbance and stress are so frequently interchanged and confused, for they are closely related. Disturbance may cause stress: conditions that result from a disturbance event may be stressful for some organisms, as might suppression of disturbance or change in disturbance frequency. Stress may also lead to disturbance: for example a great number of trees that have been killed by a prolonged drought may induce wildfire. Indeed, stress and disturbance "may act at the same time, at two or more levels, and be mutually interrelated" (Pickett et al. 1989).

Table 4.1 Common ecological disturbances and stresses

Disturbance event	Stressful condition
Earthquake	Osmotic stress
Volcano	Hypoxia/anoxia
Landslide	Desiccation
Erosion/sedimentation	Nutrient concentration extremes
Glacial movement	pH extremes
Freezing/thawing	Toxins
Coastal storm surge	Sedimentation
Wind throw	Compaction
Litterfall/senescence	Temperature extremes
Fire	Disturbance suppression
Flood	Disturbance alteration
Tide	Chronic overharvest
Upwelling	Chronic overgrazing
Nutrient flux	Pathogen
Drought	Predator removal
Herbivory	Fragmentation
Predation	Impoundment
Pathogen/parasite	Flow regulation
Invasion	
Ecosystem engineers	
Harvest	
Plowing	
Compaction	
Extraction	

Disturbance Characteristics

Disturbance events are typically recurrent in time and space, a concept that is referred to as the *disturbance regime*. Thus the frequency of occurrence, or its reciprocal the return interval, is a useful bit of information, as are the duration, spatial area, and intensity of the event. Different scholars have assigned terms to various sizes of disturbances; for example, a single treefall might be called a gap or perturbation, a patch might be an area at some stage of recovery from disturbance, and catastrophe is a term that has been used for rare events of enormous scale. For my purposes, the term disturbance will suffice for all. Another characteristic of disturbances is synergy, that is, the capacity for disturbances to occur in tandem. Drought may increase the likelihood of fire, a landslide might be brought about by volcanic activity, and a parasitic outbreak may make windthrow more severe (Rogers 1996). And, as we have seen, the effects may differ greatly by spatial scale.

In general, abiotic disturbances include events like drought, flood, tide, nutrient fluctuation, ocean storm surge, wind, landslides, mudflows, volcanoes, earthquakes, and glaciers. Biotic disturbance agents might include spikes in the activity of parasites, pathogens, or herbivores (Table 4.1). Whether or not the disturbance is desirable may depend on the management goals of the ecosystem in question. For a mid-twentieth century forester, just about any alteration to even-aged tree maturation was considered to be detrimental, and thus fire was suppressed and insects and pathogens were managed with pesticides. To the extent that disturbances can be controlled, control was the rule during this era: along with fire suppression, dams and berms kept rivers in their channels, seawalls were built to keep storm surge at bay, drought was counteracted with irrigation and flooding was minimized with drainage. In part, disturbance suppression was – and is – intended to protect human interests. But the attitude was not limited to areas of human development or commodity production. As the remaining natural areas of the country began to be protected, disturbance suppression was a primary means of ecosystem preservation. Over time, in both ecosystems managed for commodities and preserved ecosystems, the ramifications of disturbance suppression became apparent.

One reason that disturbance suppression is inadvisable is that it doesn't work, at least not forever. In fact, a period of disturbance suppression may result in a disturbance event that is much more destructive than it would have been without the suppression. The destructive fires of Yellowstone National Park in 1988 are a classic example. After decades of fire suppression, fuel in the form of living and dead vegetation built up to levels that, when ignited, burned over a greater area and with a greater intensity that the historic wildfire regime for that region. The same applies for aquatic disturbance suppression. Concrete walls and earthen berms prevent a river from flooding by preventing bankfull discharge and by facilitating the rapid flow of water downstream. When the really big rainfall comes, however, the flood is orders of magnitude larger at the point of levee breach, as the Mississippi basin floods of 1993 will attest. Neither can floodwalls eliminate storm surge in perpetuity. This painful lesson was demonstrated with Hurricane Katrina's breach of the Lake

Pontchartrain levee in New Orleans in 2005. In short, any human effort to suppress disturbance is a short term solution that will eventually fail.

An important corollary is that the probability of some disturbance events changes over successional time even without human intervention. Fire, for example, is partially dependant on fuel load, and fuel load increases as biomass accrues in the system. Pest or pathogen outbreak, the likelihood of predation or herbivory, and the action of "ecosystem engineers" also change with succession. In some cases, successional stage of the adjacent ecosystem may influence the likelihood of disturbance. For example, abundant woody debris in a riparian forest can increase flow blockage in the adjacent river and make flood more likely, whereas an early successional riparian corridor with much exposed ground makes erosion and sedimentation probable. These "internal" controls are known as system feedbacks; they are structural or functional aspects of the ecosystem of interest that influence the disturbance regime of that system. Of course, some disturbance events, such as earthquakes, ice storms, and hurricanes have origins that are completely external to the system of interest.

A second problem with disturbance suppression is that the absence of disturbance itself is an ecological stress. Disturbances are destructive, without a doubt, but in their wake they leave space for colonization and reorganized resources. To use the language of the adaptive cycle, disturbance suppression arrests the ecosystem in the Conservation Phase and prevents the Release Phase. The Release Phase is critical for r-strategists that rely upon dispersal, colonization, and rapid reproduction, but it is also essential for K-strategists that may have dominated the Conservation Phase. In any given ecosystem, many species are adapted to a disturbance regime. Serotinous conifers, fire-tolerant prairie grasses, salt marsh plants able to withstand tides, soil microbe adaptations to periodic inundation and desiccation, the correlation of mangrove development with hurricane return interval – these are all examples of disturbance-driven evolution. The abiotic environment, too, is altered by disturbance. Nutrients that were formerly sequestered in biota are suddenly released. Waste products that may have accumulated in the ecosystem are translocated to another ecosystem by burning, flooding, tidal action, and the like. Simply put, disturbance reorganizes the resources of and facilitates patch dynamics within an ecological community.

Disturbance and Diversity

The observations that disturbance suppression is stressful and that it will ultimately fail are neither new nor controversial. It has long been accepted that disturbance is a necessary part of succession, and disturbance has been used extensively as an instrument of ecosystem management. A third reason for the inclusion of disturbance in ecosystem protection is compelling but less clear – this is the hypothesis that disturbance enhances community diversity. At a basic level this is intuitive. Ecological disturbance is an agent of natural selection; it creates novel combinations of resources for colonization, it broadens the variety of available niches. Conversely, an environment that is protected from disturbance tends to become more homogenous, more completely populated by competitive dominants, and less diverse. Odum (1969) hypothesized that both the variety and evenness of organisms in an ecosystem increase as the ecosystem matures through succession, though he allowed for the possibility of a late-succession reduction in diversity due to competitive exclusion. Equally possible, again according to Odum, is that spatial stratification and the development of symbiotic relationships increase the number of available niches as an ecosystem matures. Odum's view is generally representative of the holistic climax viewpoint: that diversity and productivity are low in early successional pioneer communities, that both peak as the community enters a mature stage at which early, transitional, and late successional species are present simultaneously, and that both decline as the climax community ages (Loucks 1970).

The influence of disturbance on community diversity was presented as a general model by Connell (1978). The Intermediate Disturbance Hypothesis suggests that the alpha diversity of an ecosystem is a function of disturbance frequency. According to IDH, an infrequent disturbance regime (or the absence of disturbance altogether) would result in the scenario described above: the competitively domi-nant K-strategists would eventually exclude inferior competitors, thus limiting alpha diversity. At the other extreme, frequent disturbance would eliminate all but the best-dispersing and fastest-reproducing r-strategists. Slower growing K-strategists would have no opportunity to become established, resulting in low overall species diversity. The greatest diversity, as the IDH name implies, may be found at the intermediate disturbance frequency, which would allow the system to reach the stage of maximum life strategy representation before being reset by disturbance. In theory, the IDH can also be thought of in terms of disturbance area or intensity – again, with intermediate levels resulting in greater diversity than either extreme.

There appears to be some empirical evidence supporting IDH. For instance, when the moderate fire and grazing disturbance regimes of Mediterranean basin communi-ties are either removed or dramatically increased in intensity, species diversity declines (Rundel 1999). Even so, the IDH might lead one to think of an ecosystem as entirely within any particular stage of succession, which is misleading. In reality, every stage of the adaptive cycle is likely to be present in a given system within dif-ferent patches at different scales. Seen in this light, it is clear that the IDH is limited to diversity predictions within a particular patch at a particular scale. A between-patch model of diversity and disturbance was presented by Kolasa and Rollo in 1991. Called the Disturbance Heterogeneity Model (DHM), it predicts that habitat heterogeneity, and by association community diversity, is increased by disturbances that are smaller than the size of the community, with maximum diversity at maxi-mum heterogeneity; i.e. 50% of a given area subjected to disturbance. Extension of this model would lead one to postulate that maximum heterogeneity of the type, intensity and extent of disturbances would further increase diversity.

Denslow (1985) has suggested that disturbance is likely to enhance species diversity "when the disturbance pattern resembles that historically characteristic of the community." In other words, a disturbance regime that is similar to the regime that shaped the resident community will likely maintain that community over ecological time. A disturbance regime that is altogether different – flooding, for example, in a forest that has never experienced flood – will likely select for a community that will initially have few viable candidate species. In essence, the concept of the historically appropriate disturbance regime – tempered, perhaps, with the IDH and DMH – has become a central dogma of American ecosystem conservation and preservation. Having learned that disturbance suppression is not the way to preserve an ecosystem, we have embraced the use of a regular and moderate disturbance regime to select for the species we deem desirable. And yet, Denslow also notes that communities subjected to a rather constant physical environment and regular disturbance regime are less likely to tolerate a novel stress than a community subjected to a variable environment. So, a regular, historically appropriate disturbance regime may be a useful management scheme for the preservation and maintenance of the ideal ecosystem, but it may limit ecological response in the face of change. And change, of course, is inevitable.

The link between disturbance regime and species diversity is a critical point for ecosystem management. As we have seen, alpha diversity is linked to functional diversity and redundancy, and functional diversity is linked to ecosystem services. Humans rely upon ecosystem services and therefore require – and in some protected areas, attempt to preserve – alpha diversity. But alpha diversity in the face of future change is dependent upon the heterogeneity of the shifting patchy mosaic, and this in turn is disturbance dependent. Ecosystem protection efforts often favor the sustainable, equilibrium ecosystem that maintains its integrity over time. Is this concept compatible with the variable disturbance regime as a primary driver of diversity?

In a sense, a disturbance is an event that tests the functional diversity of an ecosystem. This is the concept of *response diversity* – the variety and range of organismal functions that are available for reaction to a disturbance (Elmqvist et al. 2003). In some cases, response to disturbance might mean recovery from disaster, as in the colonization and re-growth after a forest fire. But the disturbance need not be a destructive event – consider rainfall in a desert – and rather than recovery from disaster, the response may be better viewed as a reaction to an opportunity. Response diversity, then, is the individual and collective ability of constituent organisms to react to change. It is easy to see why response diversity should be a desirable characteristic for protected ecosystems: it allows for continued function after an unforeseen disturbance event. It is the capacity for basic ecosystem functions, like production, decomposition, and nutrient processing, to resume after temporary interruption. Of course, the responding organisms may come from within the defined boundaries of the ecosystem, or they may be new colonizers from elsewhere. It follows that ecosystems which are rich in alpha diversity or in proximity to a source of colonizing organisms would have the greatest capacity for response. As protectors of ecosystems, humans can attempt to provide the latter by protecting ecosystems in clusters or along corridors of dispersal where possible. The former, I would argue, is best achieved by the use of variable disturbance to generate spatial and temporal heterogeneity in the system of interest.

Resilience

A prevailing view of ecosystem resilience is an ecosystem's ability to retain its structure, function, and feedbacks – its ability to self-organize – after a disturbance event. Put another way, "resilience is the capacity of a system to absorb disturbance without shifting to another regime," with "another regime" being an alternate stable state (Walker and Salt 2006). Again, let us consider this concept in its strong and weak versions. At its strongest, resilience might be thought of as the ability of an ecosystem to return to its "optimal" state after disturbance, presumably in a long, stable Conservation Phase. In this sense, the disturbance is a setback to be recovered from; it is something to withstand. The "optimal" state may be considered the state of ecological health and integrity in the strong sense, consisting of native species in their characteristic numbers performing their typical functions. Recent literature defines resilience according to a property called *ascendancy*, which is a directional increase in the vigor and organization of key species in the ecosystem (Costanza and Mageau 1999; Jørgensen et al. 2007). In this view, a resilient ecosystem *ascends* to its ideal state – with the proper organisms in their normal arrangement and function – following disturbance. The terms have changed, but this is clearly the Clementsian developmental progression to the type climax. It is holism in the extreme.

How would we define resilience in the weak sense? If we separate the idea of resilience from the holistic view of the ecosystem it loses some meaning; if there was no coherent ecosystem unit before the disturbance, how can it be meaningful to expect recovery? The extreme individualist might imagine resilience to be the ability of organisms and species to colonize niches, obtain resources, and reproduce in the post-disturbance environment. There are no expectations of normalcy here. The rules of existence have simply changed, the selection pressure has been imposed, and candidate organisms will either take advantage of the opportunity or not.

The holistic preservationist might protest this last view on the grounds that some post-disturbance species are undesirable, and an assembly of undesirable organisms won't provide ecosystem services the way systems of health and integrity will. Thus, the argument might follow, it is not good enough to protect random conditions of natural selection in our ecosystems. We must strive to protect the specific conditions that are necessary to sustain the ecosystem in a desirable state. To do so, we must protect native biodiversity to maintain system integrity, regulate disturbance for the persistence of the desired community, and minimize the time and space in Release and Rapid Growth phases, during which the ecosystem is susceptible to colonization by undesirable species. This, I am suggesting, is the prevailing attitude of American ecosystem conservation.

A popular conceptual model portrays the ecosystem in a way that emphasizes the pervasiveness of the "stable unit" mindset. The ecological state of an ecosystem is commonly depicted as a ball on a two dimensional landscape, in which gravity pulls the ball downward into pits or depressions (Fig. 4.1; Peterson et al. 1998; Elmqvist et al. 2003). When it rests in a pit – or what has been called a "domain of attraction," the ecosystem is in a stable state. The width and depth of the pit represent

Resilience and multiple states

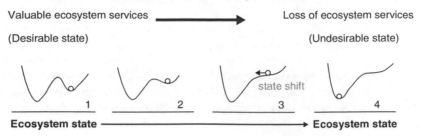

Fig. 4.1 The ball-and-pit model, as depicted by Elmqvist et al. (2003). Republished with permission of the Ecological Society of America from Response Diversity, Ecosystem Change, and Resilience, Elmqvist et al. 2003, Front. Ecol. Env., 1(9)489; permission conveyed through Copyright Clearance Center, Inc

the resilience of the stable state. In this model, a disturbance – like fire or storm – temporarily changes the conformation of the landscape, but if resilience is great enough (if the pit is deep enough), the ecosystem will recover from the disturbance and remain in the stable state. Human activities can act as stressors in the model and in effect lessen the resilience of the stable state. Overgrazing, eutrophication, over-harvest, pollution – all reduce the functional redundancy like rivets removed from the wing, and all reduce the depth of resilience on the stability landscape. In a state of weakened resilience, a disturbance can be enough to bring about a shift in the state of the ecosystem to a new stable equilibrium. The new equilibrium may be an undesirable; it may not provide the ecosystem services that the former state did. It may also be extraordinarily difficult to restore the previous ecological state if key functional groups and abiotic conditions are lacking. This model is an argument for the protection of ecosystem resilience, so that ecosystems can persist in a desirable equilibrium state in the face of disturbance.

The ball-and-pit model is of course a very simplistic representation of ecosystem change. For instance, the size and shape of the ball are fixed, so that it always responds to changes in its "landscape" the same way. The two-dimensional landscape also restricts the range of ecosystem responses to environmental change to two: it can either remain in the desirable state, or it can shift to an undesirable state. For these reasons I find the model to be a poor representation of ecosystems. An ecosystem is better represented by a globule of oil in a lava lamp (Fig. 4.2). As heat, gravity and fluid flow exert various pressures on the globule, it changes shape and size. It may combine with another globule or break into fragments. As it responds to pressures and moves through the space of the lamp, the pressures change by virtue of its position. As pressures change, the response changes. The lava lamp is still an imperfect model, but it acknowledges dynamic ecosystem response without implying that the ecosystem is an entity that resists change.

For those who acknowledge that ecosystems are dynamic but who also maintain that there are certain characteristics associated with each particular ecosystem

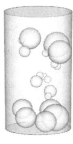

Changing stress and
disturbance over time,
with corresponding
biotic changes

Fig. 4.2 Ecosystem conformation and composition as represented by a lava lamp model. The globules of oil in the lamp represent ecosystems, which change in conformation and composition in response to environmental pressures and circumstances

type, the "range of variation" argument is a fall-back position against the individualistic onslaught. The idea is that prior to human influence, or even under the influence of indigenous cultures, ecosystems varied according to disturbance and stress regimes but that the variation was bounded by certain limits. Within these limits, the ecosystem remains recognizable – meaning that the dominant species, prevalent functions, and three-dimensional structure are more or less intact. The range of variation concept "is a recognition that complex systems, including ecosystems, have a range within which they are self-sustaining and beyond which they move into a state of disequilibium" (Egan et al. 2005). It is still an "ideal state" argument, and one that I find to be particularly slippery. In one sense it leads us right back to "strong" holistic argument – it simply broadens the boundaries of what might be considered normal. And in the end, it is still the human observer that must set the boundaries. On the other hand, if one broadens the spectrum of "normal" enough so that the "range of variation" becomes "the entire range of abiotic processes within which ecological succession is self-sustaining," then you have an individualistic argument. But I do not think that this broadest interpretation is what most "range of variation" proponents mean. Rather, I think they are allowing for some variation in what is basically the Clementsian ideal.

But as we have seen, there are problems with management for the persistence of an ideal ecosystem. The goal of protecting biodiversity is laudable, but species diversity requires a diversity of patches and therefore a diversity of disturbances in time, space, and intensity. Ecological disturbance is an appropriate and effective management tool, but regularization and homogenization of disturbance will lead to narrow response diversity. With a diversity and stochasticity of disturbance, patches of release and rapid growth will be prevalent, and the pre-disturbance assemblage may well change. The individualist sees this not as a failure of resilience but as part of the ongoing process of selection and response. Collectively, these problems suggest that "strong" holistic conservation is untenable. Ultimately, it is response diversity, spatial heterogeneity, and competition that will sustain ecological function and the ecosystem services on which we depend.

Complications of Stress

The management model I describe might be more acceptable to a holistic preservationist if all of our ecosystems existed in stress-free conditions. As it happens, they do not. Ecosystems are subjected to stress due to natural variation in climate, physical geography and biogeochemistry. As a consequence of stress, aggregate ecosystem functions become limited. In extreme drought, for example, productivity may be reduced, the rate of decomposition may decline, and nutrient cycling may be inhibited. Too much water, in the form of prolonged flooding or water-logging, may also be a natural ecosystem stressor, as oxygen availability becomes a limiting factor of function. Similarly, salinity, pH, temperature, erosion, sedimentation, and light attenuation may be considered natural ecosystem stressors. In each case, the stress limits functional capacity and alters response diversity. For instance, a certain number of species might be available for colonization and growth following a fire, but if the fire was preceded by severe drought, the number and diversity of organisms capable of recolonization might be significantly limited. Multiple stressors in combination with disturbance events may render an environment effectively uninhabitable, as there are few or no viable strategies for existence (Grime 2001).

Though many ecological stressors are natural, it seems that a great many more are the direct or indirect result of human activity. Humans have exacerbated droughts and floods, brought about anoxic conditions in inland and near shore waters, and acidified and salinated soils. Nutrient stress is a particular human forte. Many harvested environments are nutrient impoverished, while adjacent aquatic ecosystems are hypereutrophic. On other environments humans have unleashed a variety of toxins – synthetic and natural, aquatic and airborne, intentional and accidental. Humans have overharvested plants and animals, eliminated predators and pests, and overgrazed livestock on particular ecosystems. Moreover, the "natural" areas that remain are subjected to the stress of fragmentation; they are islands in a sea of agriculture, extraction, industry, development, and pavement.

On top of all this ecological stress, consider that some American ecosystems have been removed from disturbance for decades. Fire suppression until recently was the rule of forest management. Many American rivers are channelized, dammed and impounded for flood control. Sea walls minimize storm surge. Drainage tile regulates the water table on land and the flow regime in the nearby stream. In these ways humans have created ecologically stressful conditions while suppressing or regularizing disturbance regimes. Both the increased stress and the minimized disturbance contribute to diversity loss and functional limitation.

The cumulative, interactive, and synergistic effect of these various stressors may be though of as an ecosystem's *stress regime*. In many of our ecosystems, the stress regime – once defined primarily by natural constraints on growth – is now overwhelmed by anthropogenic stressors. And along with the disturbance regime, the stress regime plays a role in the selection of individual species, community assemblages, and ecosystem functions.

The American conservation community has largely responded to anthropogenic stress by protecting land from development and encroachment in varying degrees. As we have seen, the effort has been successful, as a substantial proportion of American land is under some degree of protection. It is this land, I would argue, that is often viewed as land that should be restored to emulate its former condition or preserved to maintain a desirable condition. The act of protection may remove, control, or at least mitigate some of stressors on that ecosystem – though some stressors, like fragmentation, are difficult to overcome. In recent decades, the essential role of the disturbance regime has also been recognized, such that disturbance is often used as a restoration and management tool. In many cases, however, disturbance is being used only to achieve and maintain an ideal state.

Because of the prevalence of ecological stress, organizations and individuals that are attempting to preserve ecosystems are on a management treadmill. To protect the land in question from human development is often perceived to not be enough, for in comparison with historic conditions or nearby reference ecosystems the protected ecosystem may be degraded in some way. In many cases, stressors alone or in combination have limited the presence, growth, or function of desirable organisms while creating ideal conditions for undesirable organisms. Thus, many "protected" American ecosystems have been encroached upon by invasive or non-native species. This degraded condition threatens many of the holistic ecosystem characteristics we hold dear: it lowers the ecosystem's integrity, it reduces the likelihood that the ecosystem will be resilient amidst disturbance, and it may alter some functions that are seen as components of ecosystem health. The result, in many cases, is the use of disturbance in a specific, narrow sense to remove undesirable competitors of desirable species, to create and re-create ideal conditions of the model ecosystem.

This approach to ecosystem management is compatible with the holistic school of thought and it is encouraged by national policies of ecosystem preservation. On the continuum of the weak to strong view of the ecosystem as a discrete unit, it falls far to the strong side. In such a holistic approach, it is the preservation of certain desirable species in a defined arrangement that is paramount. Resilience – the capacity of the system to recover to the desired state after disturbance – is prized. Of course, the individualist might offer the same protest we have already encountered regarding idealized ecosystems. These management efforts, the argument might go, are attempting to establish and maintain an ideal state that exists only in the minds of its proponents. Through artificial selection of species and regulation of disturbance regimes, managers are fighting against ecosystem change. As ecosystems are collections of species that are transient in space and time, this effort to achieve and maintain an ideal state will ultimately fail. Furthermore, to the extent that management efforts homogenize the system, they may hinder response to disturbance and stress in the future. In the effort to enhance ecosystem resilience, stochastic ecological response is being marginalized.

Some students of this debate propose a radical "hands off" approach, in which the land is protected from human encroachment and then left to nature's own devices. Indeed, many of our protected lands are not managed in the detailed way I describe here – they are simply too vast or too remote. And, if the capacity for disturbance in these systems is intact, the hands-off approach may be perfectly acceptable. In our smaller, more fragmented nature preserves, though, disturbance regimes have been suppressed or eliminated, and in these systems human intervention may be necessary. But what exactly is necessary? Is the re-establishment or approximation of a historic disturbance regime enough? Many of our nature preserves are subjected to numerous stressors. What if, due to stress and the prevalence of nonnative species, the protected area is of low integrity when compared with reference ecosystems? Should management efforts take steps to achieve some sort of equivalence with the reference goal? Is it appropriate for management efforts have a structural and functional goal when the ecosystem itself, on its stochastic journey through a sea of contingencies, has no goal?

References

Barrett, G. W., and Rosenberg, R. 1981. Stress Effects on Natural Ecosystems. Chichester: Wiley.

Connell, J. 1978. Diversity in tropical rain forests and coral reefs. Science 199:1302–1310.

Costanza, R., and Mageau, M. 1999. What is a healthy ecosystem? Aquatic Ecology 33:105–115.

Denslow, J. S. 1985. Disturbance-mediated coexistence of species. In The Ecology of Natural Disturbance and Patch Dynamics, ed. Pickett, S., and White, P. S., pp. 307–323. Orlando: Academic Press.

Elmqvist, T., Folke, C., Nyström, M., Peterson, G., Bengtsson, J., Walker, B., and Norberg, J. 2003. Response diversity, ecosystem change, and resilience. Frontiers in Ecology and the Environment 1:488–494.

Egan, D., Howell, E. A., and The Society for Ecological Restoration International. 2005. The Historical Ecology Handbook: A Restorationist's Guide to Reference Ecosystems. Washington: Island Press.

Grime, J. P. 2001. Plant Strategies, Vegetation Processes, and Ecosystem Properties. Chichester: Wiley.

Gould, S. J. 1989. Wonderful Life: The Burgess Shale and the Nature of History. New York: W. W. Norton.

Holland, A. 2000. Ecological integrity and the Darwinian paradigm. In Ecological Integrity: Integrating Environment, Conservation, and Health, ed. Pimentel, D., Westra, L., and Noss, R. F., pp. 45-60. Washington: Island Press.

Jørgensen, S., Bastianoni, S., Fath, B., and Muller, F. 2007. A New Ecology: Systems Perspective. Amsterdam: Elsevier Science.

Kolasa, J., and Rollo, C. D. 1991. Introduction: the heterogeneity of heterogeneity a glossary. In Ecological Heterogeneity, ed. Kolasa, J., Pickett, S. T., and Allen, T. F. H., pp. 1–23. New York: Springer.

Loucks, O. L. 1970. Evolution of diversity, efficiency, and community stability. Integrative and Comparative Biology 10:17–25.

Odum, E. P. 1969. The strategy of ecosystem development. Science 164:262–270.

Odum, E. P. 1985. Trends expected in stressed ecosystems. BioScience 35:419–422.

Odum, E. P., Finn, J., and Franz, E. 1979. Perturbation theory and the subsidy-stress gradient. BioScience 29:349–352.

Peterson, G., Allen, C., and Holling, C. 1998. Ecological resilience, biodiversity, and scale. Ecosystems 1:6–18.

Pickett, S., Kolasa, J., Armesto, J., and Collins, S. 1989. The ecological concept of disturbance and its expression at various hierarchical levels. Oikos 54:129–136.

Pickett, S., and White, P. S. 1985. The Ecology of Natural Disturbance and Patch Dynamics. Orlando: Academic Press.

Rapport, D., and Whitford, W. 1999. How ecosystems respond to stress. BioScience 49:193–203.

Rogers, P. 1996. Disturbance Ecology and Forest Management: A Review of the Literature. Forest Service Intermountain Research Station general technical report INT – GTR-336. Ogden: United States Department of Agriculture.

Rundel, P. W. 1999. Disturbance in Mediterranean-climate shrublands and woodlands. In Ecosystems of Disturbed Ground, ed. Walker, L. R., pp. 271–286. Amsterdam: Elsevier.

Sousa, W. 1984. Intertidal mosaics: patch size, propagule availability, and spatially variable patterns of succession. Ecology 65:1918–1935.

Walker, B. H., and Salt, D. 2006. Resilience Thinking: Sustaining Ecosystems and People in a Changing World. Washington: Island Press.

Walker, L. R., ed. 1999. Ecosystems of Disturbed Ground. Amsterdam: Elsevier.

Chapter 5
Invasion

What is so bad about an ecosystem that is no longer in the desirable state? What makes it undesirable? It is probably not as attractive to the human eye, for one. Who would choose a field of thistles over a field of bluestem, coneflower, blazingstar and indigo? Attractiveness in the ecological sense may also be altered; the ecosystem in the undesirable state may not support the same wildlife that it once did. A rare or endangered species that was adapted to this specific type of ecosystem may now be locally extinct. With these aesthetic and ecological departures from the ecosystem of recent history there may be a sense of lost legacy: this was the last remnant of a once-great type of ecosystem in this particular area. Aesthetic appeal and nostalgia for a particular manifestation of nature are real but subjective reasons for eschewing and lamenting ecological change. But there are also concrete, objective reasons. For instance, there may be economic ramifications to ecosystem changes that limit or eliminate hunting, fishing, or other opportunities for recreation. Perhaps the ecosystem in its new state no longer supports a particular commodity, such as timber, fish, or shellfish. There may also be quantifiable consequences of ecosystem change for a broader human audience. Maybe the change has affected some ecosystem-level functions, and maybe some of these functions could be considered ecosystem services on which humans on a local, regional, or global scale depend.

These are common reasons for the preservation of an ecosystem in a particular desirable state. Let me make two comments on the lot. First, all of the arguments stem from a change in the species complement – the biological community – of a particular region. Often, it is the invasion of one or more nonnative species that is considered the symptom of degradation, if not the disease itself. Second, all of these arguments are anthropocentric in the extreme. The new ecosystem state is undesirable, the reasoning goes, because I find its collection of species to be unattractive, or because I can no longer harvest a commodity from it, or because it is now less efficient at processing my waste. The anthropocentrism is of course understandable, for without nature's production and process we humans could not exist. In fact, I submit that all of the "undesirable state" arguments are *necessarily* anthropocentric, for without consideration for human comfort, pleasure, and well-being I see no basis for determining whether one ecosystem state is better or worse than another.

D. J. Spieles, *Protected Land*, Springer Series on Environmental Management, DOI 10.1007/978-1-4419-6813-5_5, © Springer Science+Business Media, LLC 2010

In this chapter, I explore these themes of invasion and anthropocentrism. After considering the causes, processes, and effects of species invasion as it relates to ecosystem change, I will examine the relationship of invasion with the ecosystem characteristics that humans desire: integrity, health, stability, and resilience.

Colonization and Succession

As we have seen, the colonization of species is part of the periodic reorganization that is the basis for ecological succession. In ecology textbooks, succession is typically shown as a sort of timeline, beginning with a few pioneer species becoming established in the aftermath of some disturbance event. We typically speak of colonization on a small scale, so that the initial colonizers – species that we have called r-strategists – initiate succession, arriving at and becoming established in an open patch of habitat. As depicted on the timeline, the initial pioneers are gradually replaced by longer lived, more competitive K-selected species. The later successional species gradually replace one another until, at the end of the timeline, a climax community has been achieved.

In this classic view of succession, it is not only the biological community but also the functional characteristics of the ecosystem that change along the timeline (Odum 1969). Organisms of early succession tend to be smaller and shorter lived, with simple life histories and broad physiological tolerances. As a consequence of these organismal differences, aggregate processes differ in early and late succession. An early successional ecosystem might have great photosynthetic productivity, but it has comparatively little standing biomass. It has lower diversity and less trophic complexity than a late-successional ecosystem of the same environment. In early succession, there is little biotic carbon sequestration. Nutrients are primarily extrabiotic and mobile; thus young ecosystems are nutrient-leaky. Late successional systems have a more complex array of biochemical pathways and are thus able to metabolize, process, and assimilate a greater variety of chemical compounds. Soil accretion, water holding capacity, nutrient exchange capacity, habitat complexity – all are likely to be more prevalent in late succession. The holistic view holds that late successional structure and function are self-maintaining until they are overcome by catastrophic disturbance.

Of course, no ecosystem is as neat and clean as a textbook timeline. Succession in reality is not uniform or unidirectional. Colonization is not limited to the early stages of succession but in fact occurs in all stages, as individuals of various species arrive at and exploit patches that are appropriate for their particular niche. Far from being restricted to "pioneer" status, the r-selected species are present at all stages of succession, for at all stages there are spatial and temporal opportunities for a quick bit of growth and reproduction. The "early successional" r-selected species, then, regularly exist in patches alongside "late-successional" K-selected species. This co-existence has been termed tolerance, and such species are able to tolerate one another because they use resources differently (Connell and Slatyer 1977; Tilman 1985).

To borrow Hutchinson's terminology, species are fugitives, jumping from patch to patch as chance dispersal, environmental conditions, and their own physiological requirements allow. In the broadest sense, all species may be seen as fugitives – some are just more frantic fugitives than others. Thus it would be perfectly appropriate to view the replacement, over several millennia, of a spruce forest with an oak forest as a colonization event. In any case, it is clear that ecosystems are stages for patch dynamics and species opportunism. Ecosystem function changes with the shifting patches and changing community; it will only remain stable to the extent that environmental conditions remain stable. And environmental conditions won't remain stable – particularly since the colonizing organisms themselves change their environment. We must, therefore, recognize species colonization for what it is: ubiquitous, relentless, and central to the concept of ecological succession.

Factors of Colonization

The colonization process is one of chance, with a few key probabilistic functions in the equation. The first is proximity. How close is the prospective colonizer to the patch that has suddenly become available for colonization? Naturally, closer is better, and best of all is to be present at the site of the disturbance event (and, of course, to have survived it). Plant species with seeds in the soil seed bank, for example, have a distinct advantage over plants that must arrive from afar. Disturbance events have a way of being unpredictable in space and time, so some of the best invasive species are present over a wide range and have great seedbank longevity. For species that are not present at ground zero, dispersal is the key. By wind, water, vector, locomotion, or other means of transport, the potential colonizer must arrive at the disturbed patch, where it is confronted with environmental conditions that may or may not be suitable for its survival. Can the invader germinate and grow, hatch and feed, mate and produce fertile offspring, in these conditions? Is there enough light, oxygen, moisture, nutrients, prey? The best pioneer colonizers are typically well suited to a wide range of conditions; they are stress-tolerant generalists. Even so, they face potentially intense competition for limited resources. For those species that are able to arrive and survive, the race is on. The rapid production and dispersal of numerous progeny is paramount, for this patch will soon become less suitable, and another will open somewhere else in due course.

This story depicts the colonizing species as a lucky passenger who happened to arrive at the correct stop just in time to catch the bus. To some extent this may be an accurate analogy, though it is clear that newly-established species effect certain changes in the ecosystem they inhabit. These changes are neither purposeful nor directed toward some particular successional endpoint, but they can be substantial. Examples are abundant. Invasive organisms have been shown to mineralize soil nutrients, increase or decrease nutrient availability, alter water clarity, influence microbial and invertebrate communities, and alter stream nutrient dynamics and food webs (Ehrenfeld and Scott 2001). In some cases, the changes precipitated by the colonizing species may make the patch more suitable for other species, and thus facilitate their establishment or survival. Pioneer colonizers that fix nitrogen or mobilize soil nutrients are examples of facilitators, as are species that are hosts, prey, or symbionts of other species.

The term *facilitation* correctly portrays cooperation in the case of symbioses, though much facilitation is probably less direct than it sounds. In many cases, action of the colonizing species causes a shift in the prevailing selection pressure of the patch and consequently changes the competitive landscape, making it a less hospitable place for some species and a more hospitable place for others. The opposite of facilitation may also occur; by allelopathy, herbivory, shading, nutrient sequestration, or other means, colonizers may inhibit the establishment of other species (Connell and Slatyer 1977). Through inhibition processes, the colonizer may even out-compete species that were previously dominant in the ecosystem. Facilitation and inhibition, then, are mechanisms of environmental change, meaning that change can be a product of colonization. This makes succession much more complicated than a simple series of open patches; the patches themselves are to some extent determined by the organisms that inhabit them. In other words, some colonizing organisms might be better described as drivers than passengers.

Nonnative Invaders

Before we ponder drivers and passengers, we should pause to compare the species-colonization-as-part-of-succession model as described above with the invasion of species that are not endemic to the region, which is a great concern for many who strive to protect or preserve ecosystems. Non-endemic species are known by various names – exotic, nonindigenous, nonnative, introduced, alien, adventive – but even species that are native to a particular region can grow in an undesirable way, just as some nonnative species can be desirable. To minimize confusion, I'll refer to species of aggressive growth collectively as invasives. In theory, invasives are species undergoing some sort of ecological release (Torchin and Mitchell 2004). That is, they are no longer subjected to certain aspects of environmental resistance. Typically, the term *invasive* is associated with a species that has been introduced into a new ecosystem. In its new home, it may lack the predators, parasites and pathogens that it encountered in its native range, resulting in unchecked growth and an advantage over native species. But the ecological release hypothesis has not been found to be accurate for every invasive species. In fact, there is evidence that nonnative invaders bring a significant complement of their native pathogens and that they quickly attract new pathogens, parasites and predators in their new range (Sagoff 2005). But ecological release is a real phenomenon, and it is not restricted to nonnative species. The white-tailed deer, for example, may be considered an invasive species in many areas to which it is considered native. Deer populations were formerly kept in check by wolves and cougars, and then by unregulated hunting. Released from these pressures (and aided by the human propensity to create ideal forest edge habitat), deer populations have invaded new ecosystems and become an ecological menace, checked only by cull, disease, starvation, and automobile collisions (Cote et al. 2004). From this example we can glean a basic observation about species invasion. Be it native or nonnative, a species in the right circumstances may increase its population and exploit resources, potentially at the expense of other species.

Some have hypothesized that a nonnative species may indeed differ from a native species in that the nonnative, upon release from parasites, pathogens, and predators, is free to devote resources formerly used for defense to mechanisms of competition, thereby making it a more potent invader (Blossey and Notzold 1995). This "evolution of increased competitive ability" (EICA) hypothesis is a logical extension of the ecological release concept, and there is some evidence to support it. For example, common garden experiments with Chinese tallow trees have shown that trees grown in their "new" ecosystems in the southern United States are more susceptible to herbivorous insects of its native Asian range than trees of the same species grown in the native range. Experiments with other species, however, have not shown the same effect. Both garlic mustard and St. John's wort, introduced into American ecosystems long ago, show reduced defenses against pests from their home range, but neither show increased vigor or reproductive capacity over counterparts of the same species in the native range. Furthermore, no study to date has tested the EICA hypothesis for native species experiencing ecological release in its own range.

While the mechanisms of ecological release remain somewhat ambiguous, one thing is abundantly clear: humans are extraordinarily skilled at introducing species into new ecosystems. By one estimate, 50,000 nonnative species have been introduced into ecosystems of the United States (Pimentel et al. 2000, 2001). An estimated 20% of known plant species and perhaps 10% of bird and freshwater fish species in the US are nonnative – the number of invasive arthropods and microbes is less certain. Some of these introductions are the result of intentional but ill-fated schemes. In this way, carp were introduced as a potential food source, kudzu for erosion control, Japanese honeysuckle for bird habitat, the gypsy moth for silk production, nutria for fur production, and purple loosestrife and the English sparrow for their aesthetic appeal. A great many other nonnative species have been introduced accidentally: the zebra mussel and countless others in the ballast water of ocean-going vessels, the emerald ash borer on transported wood, the white perch via canals into the Great Lakes. Still others probably found their new home as unwanted pets or garden curiosities, and so there are pythons in the Everglades, lionfish in Caribbean reefs, and Japanese barberry in the forests of New England. These are all notable invasives because they are associated with a perceived ecological harm or economic injury, and they often are the focus of intensive eradication efforts.

Not only are humans skilled introducers, but we have also cleverly devised stress regimes for our native ecosystem that make them prime recipients for invasive species. Our ecosystems are fragmented, bisected, and perforated, rich in edge habitat, and primed for colonization. Fragmentation and isolation also mean that the colonization source for the native species we desire may be far removed, across many acres of developed land. Within the ecosystem, the stress regime – including a history of disturbance suppression – can limit the potential for native species colonization in successional patches. The stress we impose is often multifaceted, involving excessive nutrient deposition, erosion or sedimentation, introduction of a toxin, overharvest, and other unfortunate side effects of human activity. The result is a somewhat impoverished biota on small scales and numerous niches available for exploitation on the larger scale. This scenario is susceptible to invasion (Levine 2000).

It is worth noting that a great many nonnative invaders are considered beneficial to humans and are therefore celebrated. All of our major crops and livestock and many of our ornamental landscape plants are nonnative, but of course it may be argued that these are not part of our natural ecosystems. What about honeybees, rainbow and brown trout, Great Lakes salmon, ladybird beetles, and earthworms? These are all examples of nonnative species that happen to be desirable. Even some of the "undesirable" nonnative species have clear ecological benefits. Zebra mussels have been credited with clarifying the waters of Lake Erie. Smooth bromegrass has proven to be an effective and drought-resistant means of erosion control. The introduced cabbage white butterfly is an important pollinator of native wild plants. Exotic species, it seems, are not always a bad thing. Conversely, there are native species that are treated as invasives – though the term *aggressive* is more often applied to native species. For example, native aspen trees are removed from prairie restorations but encouraged in nearby woodlands, native cattail and *Phragmites* reeds are deemed undesirable in wetlands, and coyotes, Canada geese, California sea lions, gulls, turkey vultures, prairie dogs, and northern squawfish have all been controlled in or removed from a portion of their native range in favor of more desirable species (Goodrich and Buskirk 1995). Yet we routinely introduce bass, channel catfish, trout, salmon, and pheasant into areas where they are not native. It all points to an attitude of ecosystem conservation that is directed by an ideal in the mind of the human beholder. Our conservation is not so much driven by a desire to protect the ecosystem in its native state as it is driven by a desire for ecosystems that are convenient, pleasing, and useful for humans.

Whether the invader is native or exotic, desirable or detestable, the central question in terms of ecosystem change is this: is the mechanism of invasive species colonization fundamentally different from the colonization of any other species in the "normal" course of succession? I suggest that it is not. Many invaders are good invaders because they are r-strategists; they reproduce rapidly and prodigiously, they disperse well, and they are stress-tolerant generalists. Happening upon a suitable patch, they do what they do best – grow and reproduce. In so doing, they may facilitate the establishment of other species and inhibit the establishment of others. In these respects the difference is only one of human perception. If the colonizing species and its associated facilitation or inhibition is desirable, it is called a pioneer; if not, it is considered an invader. What makes invader colonization appear worse than pioneer colonization is not its mechanism but its degree. First, thanks to human conveyance, the invader has the ability to be dispersed over a much greater distance, and for this reason its inhibition and facilitation effects are likely enhanced because the colonizer is more likely to be released from the population regulators in its former habitat (Torchin and Mitchell 2004). Second, the human-influenced landscape has a stress regime that favors the invader over its native competitor. Finally, the speed at which invasive colonization occurs can be staggering. The patchy mosaic of a modern American landscape is being subjected to a flurry of new invaders at an unprecedented rate.

Passengers and Drivers

This brings us back to the question of cause and effect, of passengers and drivers. Are invasive species a symptom of habitat modification and degradation, or are they purveyors of degradation? Are they instigators or only beneficiaries of ecosystem change? The question is an important one for ecosystem management, for an invasive species that displaces a historically dominant species may constitute a shift in ecosystem state. The driver-passenger debate has received some recent attention in the ecological literature, particularly with regard to invasive species expanding their range into protected ecosystems. Prevailing opinion has held that invaders can in fact be the drivers of ecosystem change; that the species themselves, by direct competition and environmental alteration, can drive native species to local extinction and thereby render irreversible ecosystem change (Vitousek et al. 1997; Didham et al. 2005). There are some clear examples, particularly among invasive pathogens, parasites, and predators. Sudden oak death, as discussed in Chap. 1, is change driven by the pathogen *Phytophthora ramorum*. Similarly, the Asian longhorned beetle is an herbivore of American hardwoods and the snakehead fish a predator; both have clearly been associated with the decline of native species. But association does not imply causation. Just because the rapid growth of the invader is associated with the decline of native population does not mean that the invader is driving the change. It may be that there are environmental stressors – like drought, disturbance suppression, or acid deposition – that are the primary cause of native species decline, thereby priming the ecosystem for a new passenger.

Speculation abounds, but empirical evidence is slim concerning the relationship of invasion and ecosystem change. In one of the few experimental analyses of the question, researchers MacDougall and Turkington (2005) tested the effects of invasive species reduction and removal in oak savannas of the Pacific Northwest. This region has been subjected to extensive invasion by exotic herbaceous species, habitat fragmentation, and fire suppression. In theory, if the invasive species are indeed drivers that are actively suppressing native species, their reduction or removal should result in a resurgence of native dominants. If, instead, native species show little or no response to invasive removal, then it seems plausible that their decline and the coincidental dominance of invasive species are both products of ecosystem stress. The results indicate that both competitive and stress-related forces are at play in this ecosystem, but the recovery of native dominants after repeated removal of invasive did not occur. This suggests that ecosystem stressors are the primary drivers of native species decline in this ecosystem, and that invasive species are merely hitchhikers on the stressed ecosystem.

I don't want to make too much of this article; though elegant in conception, it is but a single study of a small area over a limited time. And there are too many examples of invasive species that clearly out-compete or flat out kill native species to subscribe completely to the passenger model. Clearly, this is another example that wonderful ecological mantra: it is neither one extreme nor the other. Still, MacDougall and Turkington provide evidence that invasive species are no different than native species in that their capacity for colonization is a function of, and a response to, environmental conditions. Given the right circumstances, by opportunistic resource use, superior stress tolerance, or some combination of the two, they may indeed supplant the previous dominant species and shift the state of the ecosystem. Anthropogenic ecological stress, in this and many other cases, is obviously advantageous for the invader. If there is a lesson here for ecosystem protection and management, it is that the war of eradication on invasive species is futile if the ecosystem stressors remain intact.

But there is a larger point here, one concerning successional change, ecological function, and ecosystem health and integrity. Given that species colonization is part and parcel of ecological succession, and further that aggregate ecosystem function is specific to the species complement and the environmental context, how can the maintenance of a particular ecosystem function mean anything but the maintenance of a particular species list? And if ecological integrity and ecosystem health are really tied to a particular dominant species community, how does this differ from the concept of the climax community? On the other hand, if the species complement may change but integrity and health remain intact, who is to say that the function of any complement of species is not perfectly acceptable?

I am not making the claim that the introduction of invasive species is not a problem. Clearly it is, but I would argue that it is a problem for humans, not for ecosystems. It is a problem for the human view of an aesthetically pleasing landscape, for the preservation of a cherished species or a perceived ecological legacy. It is a problem for the production and harvest of commodities that are subject to injury by the invader. But I am hard pressed to see how the invader damages the ecosystem. True, the components of the system will change, and so then will the processes that link the components. But absent the human imperative, how can one set of processes and functions be considered better or worse than another?

Ecosystem Function and Service

Having presented the case that (1) invasion and colonization are part of succession; (2) invasion is partially a function of stress and disturbance regimes; and (3) the colonization of native species and the invasion of nonnatives are fundamentally the same process, I now turn to some questions of ecosystem service. Specifically, what services do we want or need our ecosystems to provide? Is true that these services require the system to be maintained in an ideal state? And, most critically, does invasion threaten ecosystem service?

The Millennium Ecosystem Assessment (2005) identified four categories of ecosystem services on which humans depend. One category is defined as *provisioning* services. These are products, like food, water, pharmaceuticals, and materials that ecosystems provide for humans. A second category of services is called *regulating*; these are processes like decomposition, water and air purification, sequestration, pollination, and disease control. *Supporting* services are essential but less direct, including things like seed dispersal and nutrient redistribution. Finally, *cultural* services provide humans with inspiration, recreation, and opportunities for discovery. Now, my question is this: are these services associated with a particular ecosystem state? The answer depends upon the category. Cultural services, it seems, may well be dependent upon an ideal ecosystem state, the loss of which would mean the loss of significance or inspiration. Similarly, some food products and ecological commodities are found only in a particular ecosystem states. On the contrary, regulating services would seem to require ecosystems at various stages in order to process a wide variety of biochemical material over time. Regulating services are often complementary. For instance, we need ecosystems to fix nitrogen from a gas to ammonium and also to convert it from ammonium back to a gas. We need production as well as decomposition, water retention as well as aquifer recharge, sequestration as well as mobilization; no single state of any ecosystem can provide all the regulating services. Supporting services are likewise scattered among all successional stages, though some, like nutrient and seed distribution, are logically associated with early successional stages. And so, some essential ecosystem functions are dependent upon the ecosystem maintained in a stable, ideal state, while other functions are not; in fact some functions are enhanced by the shifting mosaic of patchy succession.

We have seen this discordance before, in another guise. It is the last century of debate in American ecosystem protection writ small. The cultural and provisioning services are apparent in John Muir's mission to defend and preserve the inspirational aesthetic of nature and in Gifford Pinchot's quest for conservation in the name of harvest. The regulating and supporting services are ecosystem processes that are more compatible with the views of functional redundancy and interchangeability; one may find the seeds of these concepts in the work of Eugenius Warming. The dichotomy of purpose, traced through the work of Gleason, Hutchinson, Leopold, and many others, is apparent in the dilemmas of ecosystem management today.

Ecosystem Services and Invasion

And which of these services, being dependent on the ecosystem in its ideal state, are threatened by invasive species? Clearly, it is culture- and commodity-based preservation that must strive to maintain stability by excluding undesirable species. Invasion of European buckthorn into Redwood national forest would not do at all. Purple loosestrife reduces waterfowl habitat and hence affects hunting, while myxozoan parasites damage trout fisheries. The sea lamprey, zebra mussel, and cogongrass are three of many, many invasive species that have already resulted in millions of dollars of lost aquatic and forest commodities. The impact of nonnative species on supporting and regulating services is not as clear. Cattails have encroached into the sawgrass of the Everglades; this is unsightly, perhaps, but is it detrimental to the biogeochemical processing and distribution of the ecosystem? Cattail plants fix carbon and transpire water just as sawgrass does. They contribute organic matter to the soil, sequester contaminants, and redistribute nutrients. They provide habitat for organisms. Do they have the same metabolic rates, wildlife value, and life history as sawgrass? Certainly not, but is that a necessity?

Perhaps cattails are too close to being native and too close to the structure of sawgrass to be an effective example. In cases where the invader is a more novel addition to an ecosystem, functional alteration is expected to a greater degree. So let's consider another example (Toft et al. 2003). Water hyacinth is native to Brazil and is one of the most invasive and fastest growing plants in the world. It has invaded North American waters, and in many ways it has become a driver of ecosystem change. In a study of the functional effect of water hyacinth invasion in California, the invasive plant was found to cut light penetration and reduce the dissolved oxygen levels to a far greater extent than the pennywort, a similar native floating aquatic plant. Water hyacinth supported a different invertebrate assemblage than the pennywort, and it was less useful for native fish populations. Clearly, this invader significantly changed the ecosystem despite its morphological similarity to the native plant – it is not functionally equivalent to the species it displaces. But the hyacinth did support an invertebrate and fish community, and at some sites the faunal community of the hyacinth was more dense and diverse than the pennywort community. Human preferences aside, how can it be said that the water hyacinth-dominated ecosystem is of lower quality than the native ecosystem? Couldn't this be viewed as an aggressive pioneer with inhibitory tendencies?

It is not my intention to be the great defender of water hyacinth, or the champion of the zebra mussel, or the defense counsel for kudzu. It is obvious that these and other species are causing great economic harm to human societies, and it is equally obvious that human societies have brought this upon themselves through careless transport, introductions gone awry, and ubiquitous ecological stress. But to extend this human-centered harm into a broad condemnation of all species that leave their native range and invade another? This is a condemnation of succession itself. The fact that so much ecosystem management is focused on the control of nonnative species demonstrates that the human need for stable, native ecosystems – ecosystems of health and integrity; ecosystems that return to the same state after disturbance – is at odds with the individualistic, exploitative, shifting mosaic of ecological systems.

Mark Sagoff (2005) has made the case that the concept of ecological harm caused by invasive species is nebulous at best and downright unscientific at worst. In his estimation there simply is no clearly defined set of criteria that distinguish the nonnative invader from the native opportunist. Furthermore, Sagoff cites numerous studies that have experimentally shown no difference in the propensity for ecosystem dominance between native and nonnative species. In fact, Sagoff argues that there is evidence for a positive association of nonnative species and ecosystem species richness. Here again, we cannot equate association with causation; species richness may not be the result of invasion, it may simply be that systems ripe for invasion also have qualities that encourage general species richness. But Sagoff's point is that, far from being universally associated with ecological harm, nonnative species are often associated with increased diversity. Daniel Simberloff (2005) has offered a sharp rebuttal, arguing along numerous lines that nonnative species are indeed harmful to natural environments. But on one point, at least, Sagoff and Simberloff appear to agree: concepts of ecosystem health, integrity or stability that exclude the presence of nonnative species by definition are tautological and problematic. The detrimental effect of nonnative species on health, integrity, or stability is hard to define if one cannot clearly define health, integrity, or stability to begin with.

Eradication

American reaction to the introduction of nonnative species, once one of approval and encouragement, has in recent decades become a kind of selective warfare. All manner of chemical, mechanical, and biological weaponry, along with inspection and quarantine, have been employed to check the influx of nonnative species and to control those that are already established. To be sure, the nation's systems of agriculture, horticulture, silviculture, and aquaculture are dependent on pest control. We can control pests over a small space and short time, but on the grand scale our efforts have been an abysmal failure. Due to new introductions, our reliance on chemical control, and the resultant artificial selection, the number of pesticide-resistant pests has increased tenfold in the United States since 1950. Despite quarantine and inspection, a new nonnative species becomes part of the San Francisco Bay ecosystem every few months, and thousands of nonnative plant propagules are introduced to the St. Lawrence Seaway each year (Georghiou 1990; Solow and Costello 2004; Cohen et al. 2007). And these are just the ones we detect. Furthermore, I am unable to name a single nonnative species that has become established, experienced explosive growth, and then been successfully eradicated from the American landscape. Once they arrive, they are part our ecosystems, like it or not.

In our efforts to protect, restore, preserve or conserve ecosystems that are not intended for commodity production or harvest, invasive species control is a standard tool in the toolbox. Ecosystem managers are burning buckthorn with blowtorches, killing nonnative (and native) fish with rotenone, treating wide swaths of land with herbicide, and using all manner of burning, girdling, addling, cutting, netting, and shocking to give desirable species a selective advantage over undesirable species. Since these ecosystems are not providing provisional services, and since there is no evidence that the nonnative species uniformly result in dysfunctional ecosystems (and some evidence that certain functions are maintained or enhanced with invasion), I can only conclude that we mount such an offensive out of regard for legacy.

A Conservation Dilemma

When I ask my students why we wage war on invasive species in non-provisional settings, I commonly get two related responses. The most common response is based on our ecological legacy: that we (meaning European Americans) inherited particular ecosystems and species upon arrival to this continent, and we are failing in our duty to maintain those systems. We must, therefore, do what we can to restore, protect and preserve the remnants of our nation's ecosystems in their historic state. It is the "leave no trace" philosophy on a continental scale. The second common response is some representation of the precautionary principle. In essence: if a nonnative species might cause irreversible harm to an ecosystem, and if we lack evidence or knowledge that it will *not* cause harm, then the species should be excluded pending such evidence.

I understand and empathize with both responses. But let's consider where these attitudes lie on our conservation continuum (Fig. 5.1). They advocate the protection of a discrete set of desirable species – the species that were present at the time of European American contact. They show a preference for the preservation of an ecosystem in a stable state at some particular phase of succession. They favor a return to the ideal stable state and characteristic function following disturbance and a contempt for species that might alter processes and functions associated with the ideal state. These are concepts of idealistic holism, and they bear the mark of the four horsemen of ecosystem preservation: integrity, health, stability, and resilience. Billed as the "sustainable ecosystem," this is an expectation that can never be met, and there may be great harm and incredible cost in the attempt.

And how might we view invasive species in a more individualistic, non-equilibrium, stochastic, radically contingent form of ecosystem protection? There might be less emphasis on the control or eradication of undesirable species, for these might be considered wasted efforts if new introductions continually occur and if the stress regime we place on our ecosystems is not reduced. Rather than the preservation of particular desirable species in a characteristic arrangement, the emphasis might favor the reestablishment of a disturbance regime and the treatment of that regime as a force of open selection, not as a hindrance to "normal" succession. A goal of ecosystem

**Disturbance and Invasion
Conceptual Continuum**

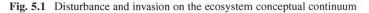

Radically Individualistic

- Disturbance as an alteration of prevailing stress regime
- Disturbance as an opportunity for colonization
- Colonization of new species as a response to opportunity

Strongly Holistic

- Disturbance as a departure from the ideal—to be resisted
- Prescribed disturbance for maintaining the ideal
- Colonization of new species inherently undesirable

Fig. 5.1 Disturbance and invasion on the ecosystem conceptual continuum

protection in this view might feature the reduction of anthropogenic stress and the maintenance of opportunities for species distribution, migration, and competition.

This is the dilemma of ecosystem conservation. We humans have real and perceived needs to define and maintain stable ecosystems, but the species that populate ecosystems are driven by contingent response to environmental fluctuation. If the boundaries that we assign to our ecosystems are imaginary, the lists of suitable species transient, and the concept of appropriate function contrived, how are we to proceed? What exactly should occur on this land that we protect?

To this point, I have considered the origin, development, and structure of the holistic and individualistic concepts of the ecosystem. I have shown that the definitions of ecosystem attributes are not at all clearly delineated into two paradigms; the reality is that current ecological opinion falls along a range between the extremes. But it is also evident that American ecosystem management policy is predicated on the principles of holism, even though the tenets of holism – stability, health, integrity and resilience – do not stand up well to nonequilibrium science. To evaluate the ramifications of this conceptual dilemma, we now turn to an analysis of ecosystem management in practice.

References

Blossey, B., and Notzold, R. 1995. Evolution of increased competitive ability in invasive nonindigenous plants: a hypothesis. Journal of Ecology 83:887–889.

Cohen, J., Mirotchnick, N., and Leung, B. 2007. Thousands introduced annually: the aquarium pathway for non-indigenous plants to the St Lawrence Seaway. Frontiers in Ecology and the Environment 5:528–532.

Connell, J., and Slatyer, R. 1977. Mechanisms of succession in natural communities and their role in community stability and organization. The American Naturalist 111:1119–1144.

Cote, S. D., Rooney, T. P., Tremblay, J. P., Dussault, C., and Waller, D. M. 2004. Ecological impacts of deer overabundance. Annual Review of Ecology Evolution and Systematics 35:113–147.

Didham, R. K., Tylianakis, J. M., Hutchison, M. A., Ewers, R. M., and Gemmell, N. J. 2005. Are invasive species the drivers of ecological change? Trends in Ecology & Evolution 20:470–474.

Ehrenfeld, J., and Scott, N. 2001. Invasive species and the soil: effects on organisms and ecosystem processes. Ecological Applications 11:1259–1260.

Georghiou, G. P. 1990. Overview of insecticide resistance. In Managing Resistance to Agrochemicals: From Fundamental Research to Practical Strategies, ed. Green, M. B., LeBaron, H. M., and Moberg, W. K., pp. 18–41. Washington: American Chemical Society.

Goodrich, J., and Buskirk, S. 1995. Control of abundant native vertebrates for conservation of endangered species. Conservation Biology 9:357–1364.

Levine, J. 2000. Species diversity and biological invasions: relating local process to community pattern. Science 288:852–854.

MacDougall, A., and Turkington, R. 2005. Are invasive species the drivers or passengers of change in degraded ecosystems? Ecology 86:42–55.

Millennium Ecosystem Assessment. 2005. Ecosystems and Human Well-being: Synthesis. Washington: Island Press.

Odum, E. P. 1969. The strategy of ecosystem development. Science 164:262–270.

Pimentel, D., Lach, L., Zuniga, R., and Morrison, D. 2000. Environmental and economic costs of nonindigenous species in the United States. BioScience 50:53–65.

Pimentel, D., McNair, S., Janecka, J., Wightman, J., Simmonds, C., O'Connell, C., Wong, E., Russel, L., Zern, J., and Aquino, T. 2001. Economic and environmental threats of alien plant, animal, and microbe invasions. Agriculture, Ecosystems and Environment 84:1–20.

Sagoff, M. 2005. Do non-native species threaten the natural environment? Journal of Agricultural and Environmental Ethics 18:215–236.

Simberloff, D. 2005. Non-native species do threaten the natural environment! Journal of Agricultural and Environmental Ethics 18:595–607.

Solow, A., and Costello, C. 2004. Estimating the rate of species introductions from the discovery record. Ecology 85:1822–1825.

Tilman, D. 1985. The resource-ratio hypothesis of plant succession. American Naturalist 125:827.

Toft, J., Simenstad, C., Cordell, J., and Grimaldo, L. 2003. The effects of introduced water hyacinth on habitat structure, invertebrate assemblages, and fish diets. Estuaries and Coasts 26:746–758.

Torchin, M., and Mitchell, C. 2004. Parasites, pathogens, and invasions by plants and animals. Frontiers in Ecology and the Environment 2:183–190.

Vitousek, P. M., Mooney, H. A., Lubchenco, J., and Melillo, J. M. 1997. Human domination of Earth's ecosystems. Science 277:494–499.

Part II
Ecosystems in Practice

Chapter 6
Very Small Ecosystems

A common criticism of the individualistic, nonequilibrium view of ever-changing ecosystems is that "natural" ecological change happens over great time scales – so great, in fact, that such change is irrelevant for our current ecosystem preservation and conservation efforts. Of course ecosystems change, the argument goes, but they change over millennia. The holistic equilibrium view is based on the scale of years to decades, and on this scale ecosystems may be treated as stable – progressing through succession to the domain of attraction – and would remain stable if it were not for human activities. Without a doubt, humans have accelerated ecosystem change by creating stressful ecological conditions and by introducing and unintentionally favoring invasive species. It is also clear that the dominant plant and animal communities of an ecosystem, given a regular disturbance regime and a constant stress regime, can remain relatively unchanged over time. Indeed, ecological changes on the scale of years to decades may be subtle; individual species may a respond to a gradual increase in regional temperature, perhaps, or slow processes of erosion and sedimentation. As we have seen, such responses can result in monumental ecosystem change over long periods of time, but it is true enough that these changes may not even be noticeable on an annual basis. Therefore, the argument may conclude, it is our duty to restore ecosystems to and maintain them in the appropriate stable state.

A counter-argument may also be made on the basis of scale. The holistic, ideal ecosystem is stable and at equilibrium only on the temporal and spatial scale that is convenient for the human experience. The collections of species that we consider to have integrity, the ecosystem functions that we consider to be healthy, the response to disturbance that we consider to be resilient – all are products of chance that coincide with our own scale of perception. By evolutionary analogy, the "domain of ecosystem attraction" is only stable in the same sense that species appear to be immutable. And so, we may consider a certain ecosystem arrangement to be a part of our legacy, we may find it to be inspirational and culturally significant, and we may see that it provides humans with valuable products or functions. But it is quite another thing to say that this perceived stable state is what the ecosystem *should* be.

The case for the individualistic view of ecosystems over large spatial scales and long time frames has been demonstrated elegantly by paleoecologists. The following chapters of this book are devoted to the consideration of ecosystems on scales of

D. J. Spieles, *Protected Land*, Springer Series on Environmental Management,
DOI 10.1007/978-1-4419-6813-5_6, © Springer Science+Business Media, LLC 2010

time and space that are more relevant to humans. First, though, I wish to bracket the paleoecological evidence for individualistic ecosystems with evidence from the small-scale ecosystems of the microbial world.

Microbial Ecosystems

Ecologists have often noted that we know very little about the species with which we share this planet. About two million species have been identified and named by taxonomists; the actual total may be anywhere between four and 100 million species (Wilson 2002). This alone is a daunting thought, but consider that identification work to date has preferentially focused on animals and plants. Comparatively, almost nothing is known about microorganisms. Of the identified two million species, only a few thousand are species of bacteria, and about 100,000 are fungi and algae. And yet it is microorganisms that truly dominate the planet. By one estimate, there are about 5×10^{30} living prokaryotic cells on earth (Whitman et al. 1998). The number of species to which these cells belong is "widely held to be unknown and unknowable," and in fact the whole species concept, which is largely based on sexual reproduction, is problematic for microorganisms (Curtis et al. 2002). Even so, bacterial diversity can be estimated based on the variety of nucleic acids extracted from a particular environment. This can give us a ballpark approximation of what must be a staggering planetary diversity: 160 bacterial species per milliliter of seawater, and approximately 20,000 species per gram of soil. These rough estimates give rise to more questions about the unseen hoards: "Who are they?, and What are they doing?" (Curtis et al. 2002; Ward 2002).

In certain respects, the microbial ecosystem is analogous to macroscopic ecosystems. There are producers, both photoautotrophs that use solar energy and chemoautotrophs that use chemical energy to fix carbon. There are consumers – chemoheterotrophs, which must ingest and obtain energy from carbon like we humans, and photoheterotrophs that metabolize carbon with light as an energy source. There are predators, parasites, pathogens, scavengers, and symbionts. There are even life strategies that correspond with r- and K-strategists: those that grow explosively when nutrients are plentiful and conditions optimal and revert to periods of latency or dormancy when conditions are less than ideal, and others that are superior competitors of slow, steady growth in a low-nutrient environment. Once established, the various species compete fiercely for resources. There are stressors in the microbial world; some, like desiccation, hypoxia, and osmotic stress are the same factors that stress macroscopic ecosystems. There are also disturbances. Sudden inundation, rapid nutrient influx, or agitation might disrupt the microbial ecosystem and send it into a reorganization phase. So the components of an ecosystem are all here. But do microorganisms behave as a community the way we understand the communities of forests, lakes and coral reefs?

Indeed they do. While it is true that microorganisms can and do occur singularly, it is now becoming clear that prokaryotic organisms predominantly exist in communities that respond to and effect changes upon their environment. In a word, they are ecosystems. Commonly called biofilms, these microbial communities are recognizable to humans as the slime that grows on slippery rocks in a stream, on a long-submerged boat hull, or on a contact lens. The ability of microbes to form such communities is an ancient and widespread survival mechanism for prokaryotic life on earth (Hall-Stoodley et al. 2004). Microbial colonization and growth is reminiscent of macroscopic ecosystem succession (Fig. 6.1).

Biofilm colonizers are individual bacteria, fungi, protists, or algae that, while drifting through their aquatic environment, collide with a solid surface with which they share a weak molecular attraction. These are the pioneer species of the biofilm ecosystem. Successful colonizers are able to secure their purchase with sticky, hair-like appendages and then by secreting a kind of biological glue to secure their hold amidst the flotsam of the microscopic world. This may include cellular debris, inorganic material, and even other organisms, all of which may or may not become attached to the fledgling community. Just like windblown seeds that land in soil, germinate, and begin the struggle for life, colonizing microorganisms compete for space and resources. One way in which microbial pioneers protect their advantage is by secreting prodigious amounts of an extracellular polymer that forms the physical barrier of a slime coat around their point of attachment. But the defenses are not only physical. Bacteria in marine biofilms have been shown to produce a potent toxin which paralyzes and kills the amoeba that otherwise threaten to consume them (Matz et al. 2008). If this were succession in an abandoned field, we would call such action inhibition.

Facilitation also occurs in biofilms, though it is more commonly known as *recruitment*. The established colonizers provide attachment sites that enable other microbes to join the growing colony. While some organisms can apparently attach

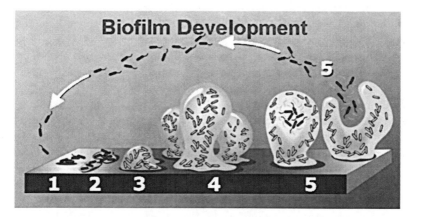

Fig. 6.1 Successional development of a biofilm community. Stage 1: initial attachment of cells to the surface. Stage 2: production of extracellular polymeric substances resulting in more firmly adhered "irreversible" attachment. Stage 3: early development of biofilm architecture. Stage 4: maturation of biofilm architecture. Stage 5: dispersion of single cells from the biofilm. Republished with permission of Annual Reviews, Inc. from Biofilms as complex differentiated communities by Stoodley et al. 2002; permission conveyed through Copyright Clearance Center, Inc

themselves to the extracellular slime coat, others attach to specific binding sites provided by the pioneer organisms. By both inhibition and recruitment, then, the pioneer organisms set the stage for biofilm development and growth. Maturation of the community is a function of the physicochemical environment, but there is clear evidence for abundant cell-to-cell communication (Molin et al. 2000). Communication may play a role in the structural formation of the community and in resource use, for constituent organisms of the biofilm are to a certain extent interdependent. Organisms with similar functions – sulfur reduction, for example – might constitute a guild. The biofilm is a collection of guilds, much as we consider a macroscopic ecosystem to be comprised of trophic guilds. For certain functions, guilds in a biofilm community are syntrophic with other guilds, meaning that each guild metabolizes the waste products of another. Thus the microbial community can be a series of interlocking syntrophic relationships, such that a given guild is exchanging resources and products with several other guilds and with the surface itself. Altogether it is a picture of an adapted and highly organized biofilm assemblage.

The organization even extends to three-dimensional structure, for as the community grows it develops into a mushroom-like matrix that is not uniform within, but rather comprised of aggregates of cells – patches, one might say – interspersed with channels for water flow. According to Mary Ellen Davey and George O'toole (2000), "Numerous conditions, such as surface and interface properties, nutrient availability, the composition of the microbial community, and hydrodynamics, can affect biofilm structure." Biofilm structure is affected by perturbations in the environment, such as flow rate and turbulence and changes in nutrient availability, but it is partially driven by the microorganisms themselves. The constituent organisms' range of motility, access to particular resources, and density all play a role in biofilm architecture. Through chemical quorum-sensing systems, the rate of growth and differentiation is regulated and the biofilm structure takes shape (Karatan and Watnick 2009). In this way the community grows in complexity and size, and as its patches differentiate into interdependent guilds it develops what might be thought of as a set of aggregate functions. Some biofilm functions are extraordinarily useful to humans while others are a scourge, but of course from the microbial perspective, humans might be a source of nutrients, a point of attachment, or irrelevant altogether.

Biofilm communities are surprisingly beautiful in arrangement and function. They are also impermanent. After a certain period of growth through recruitment and cell division, guild differentiation, and development of mature synergistic (and perhaps antagonistic) function, the biofilm community reaches a stage of dispersion. Driven by some combination of environmental cues and internal communication, organisms of the biofilm community secrete enzymes that degrade the slime-coat matrix. The microbes are then released. By individual motility or as clumped aggregates swept away by flowing water, they leave the former biofilm structure (Hall-Stoodley et al. 2004). Upon release, some cells may perish by predator or parasite. Others may be swept away to die in an unfortunate encounter with inhospitable conditions. Some, though, disperse and find new points of colonization. These are sites of new biofilm ecosystems – just as the site that they vacated is a site for the potential colonization of new pioneers. And all of this, from pioneer colonization through maturation and dispersion to colonization again, may take place over the span of a few hundred micrometers and a few days, weeks, or months (Picioreanu et al. 2000).

Biofilms as Model Ecosystems

These collections of organized and specialized cells, capable of chemical communication and interspersed with a network of circulating water, bear obvious resemblance to multicellular organisms (Molin et al. 2000). But these are not multicellular organisms, they are communities. Highly organized though it is, the biofilm is not populated only by compatible species performing their functions in harmony. Julian Wimpenny (2000) has suggested that the biofilm is more like an ecosystem than a tissue, and as such the "microbial community will consist of a mélange of types. These will include primary resource converters; secondary and subsequent species relying on products of a food chain; scavengers that do not contribute to the efficiency of the community or may even detract from it; parasites, predators, and competitors, none of which represent added value for the association. What is more, as time goes by, other species will be imported or exported so that the community will change in ways that may or may not be energetically favourable." In fact, the biofilm-as-superorganism hypothesis has not been supported by experimental evidence. For instance, organisms that form a coherent community in one nutrient environment have been shown to separate when subjected to a different condition. Even biofilms consisting of a single species are not fixed to a particular structural organization; rather, the structure is a complex result of environmental condition and organismal response (Molin et al. 2000).

Certainly, biofilms and macroscopic ecosystems are not perfectly alike. Forests and prairies do not have a stalk-like matrix of slime holding them together, nor do they fly apart at some chemical signal. Likewise, biofilms do not exhibit the same sort of gradual interspersion with one another that we see in larger ecosystems. Nonetheless, the similarities are striking. A biofilm's ecosystem-like attributes and rapid pace of succession makes it a potentially useful model for the evaluation of our ecosystem concepts. Do integrity, health, resilience, and stability make any sense for biofilms? Stability would naturally have to be considered on a much smaller scale; even so, it is clearly ephemeral in microbial communities. Some biofilms are apparently remarkably persistent amidst treatment of detergents and antibiotics, but stability suggests not only persistence but also constancy of structure and function (Molin et al. 2000). Biofilm structure changes with maturation; species and individuals arrive and depart; the relative density of species changes as the community grows; predominant function changes with resource availability. Furthermore, biofilm systems regularly collapse and disperse. None of these characteristics would seem to indicate inherent stability. But as we have seen, stability is a relative concept. If we choose to define the system as the interaction of symbiotic organisms over a discrete period, I have no doubt that we could find instances of stability.

How about ecosystem health and integrity? The terms we use them to describe macro-ecosystems can really only be understood in terms of a native reference system. For biofilms, and really microorganisms in general, we have no such understanding. It is possible, though, for a sort of invasion to happen in biofilm communities. For example, despite physical and chemical barriers to invasion, both viruses and bacteria have been shown to invade with the potential to alter the structure and function of the biofilm community. It is also possible that a shift in environmental conditions, like hydrodynamics or nutrient availability, can change the selection pressure and consequently alter the structure of the biofilm. As a consequence of environmental conditions and genetic responses, biofilms occur in a variety of structures, ranging from gangly, spatially diverse bulbs to flat homogenous structures – each unique structure presumably with its own functional character (Doolittle et al. 1995; Burmolle et al. 2006). So invasion and environmental variation can alter the biofilm community, but we have no basis for determining whether one biofilm structure has more integrity than another, or whether one function is healthier.

Actually, we do have a basis for judging the health and integrity of some biofilm systems – these are the microbial ecosystems that we actively manage. Two examples will illustrate the point. We use biofilm communities in the treatment of our wastewater, primarily to convert solid organic waste to gases like carbon dioxide and methane. These communities have health and integrity when their metabolic and reproductive rates are high and steady. Erratic substrate, inadequate oxygen, or toxins like bleach in the waste stream can all reduce the health and integrity of this type of biofilm. I also actively manage the biofilm on my teeth by brushing, flossing, and trying to maintain a reasonably healthy diet. Should I neglect any of these for an extended period, my unhealthy oral biofilm would soon be obvious to both my wife and my dentist. The point is this: like macroscopic ecosystems, the only way we can evaluate the integrity and health of biofilm ecosystems is if, by some function or appearance, they are of benefit to humans.

How about resilience? For an ecosystem to be resilient, as we have seen, it must maintain its basic structure and function after disturbance. In the case of biofilm ecosystems there might be small perturbations during the maturation process – causing a few cells to be sheared off or killed here and there – but dispersion is the major disturbance event. Upon dissolution of the matrix, the biofilm organisms, nutrient gradients, chemical signaling networks, and flow channels all dissipate; this would appear to be parallel to the destruction of biomass structure and organization in the Release Phase of the adaptive cycle. To be resilient, then, the biofilm components would be expected to reform communities of similar structure and function at a new point of colonization. Put another way, we might ask: is the post-dispersion reorganization of structure and function predictable?

This is a difficult question to answer with precision, for biofilm structure can vary in ways subtle to the human observer but critical to the microbial world. The geometric structure is important, but so are the concentration and variety of solutes, the types and abundance of species and their distribution, arrangement, and activity, and the density, permeability, and viscosity of the matrix (Picioreanu et al. 2000).

To some degree, these characteristics are controlled by the genetic expression of the colonizing organisms themselves: they are known as intrinsic factors. In theory, a complete understanding of the genome of each constituent organism would provide us with a limited predictive power for ecosystem assembly. At present, we lack such a complete understanding. Even if we had it, we would still be poor predictors without knowledge of the extrinsic factors of the microbial environment. Extrinsic factors might even be less knowable that intrinsic; they are the stochastic fluctuations of the microclimate (Wimpenny 2000).

Not surprisingly, microbial response to environmental change is still poorly understood, and so is the mechanism of biofilm assembly. The best attempts at understanding biofilm geometric structure have been simulations that account for the effects of extrinsic factors like hydrodynamics, substrate form and solute concentration on growth, attachment, and detachment. Assuming that microbes would respond to these variables in a uniform and consistent manner (which in real life they would not), one might simulate the growth biofilm ecosystems over space and time to test the hypothesis that biofilm ecosystems are structurally resilient. In fact, the simulations show quite the opposite – biofilm growth has no fidelity to geometric structure over space or time. Instead, the structures of post-dispersion biofilm communities are strongly influenced by micro-variation in the physical and chemical environment (Picioreanu et al. 2000). True, these are only simulations of extrinsic factors. And of course, the "real world" response to environmental conditions is driven in a specific way by the organisms of the biofilm ecosystem. But assuming that the colonizing organisms in each reorganization may well be different in type, number, and arrangement, the intrinsic effect would make it even less likely that biofilm ecosystems are resilient and mature in the same way time after time. On the contrary, it's all contingent upon the peculiarities of place, moment, and constituent organisms.

Perhaps the minute details of biofilm structure are unimportant for resilience. Maybe resilience should be framed in broader terms of function. Does a biofilm community that functions by metabolizing ammonium to nitrous oxide, upon dispersion, re-form communities that also metabolize ammonium to nitrous oxide? Do the biofilm communities on my teeth spawn similar communities that eat away at my enamel and taint my breath? If they do, then biofilm ecosystems are clearly resilient; but if this is all we mean by resilience, then it is definitely a property of macroscopic ecosystems as well. A native forest replaced by nonnative shrubs and vines still does photosynthesis and respiration. An algae-choked lake still has trophic function. No, this is not what we mean by resilience in our macroscopic ecosystems. For an ecosystem to be resilient in the "strong" sense it must be some reincarnation of the species list and three-dimensional structure of the pre-disturbance state. If biofilm ecosystems are a useful model, it would seem that "strong" resilience is only likely under the most stable environmental conditions.

Such stable conditions may exist at certain places on earth. There are, for example, microbial communities in the water-filled pore spaces of rock deep below the planet's surface. They are known as SLiMEs – subsurface lithoautotrophic microbial ecosystems. The environment is extremely nutrient poor, and consequently the metabolic rates of SLiMEs are among the lowest ever recorded (Stevens and McKinley 1995).

The ecosystem appears to be based entirely on geochemically produced hydrogen, without access to any product of photosynthesis, past or present. Relatively protected from fluctuations of climate, nutrient availability, temperature, pressure, and flow rate, these must be some of the most stable ecosystems in existence. Though SLiME succession has not been studied, I can imagine a dispersion-reorganization scenario that is remarkably resilient. I also imagine that such a stable environment is the exception to the rule of earth's ecosystems.

Biofilms as Patches

Biofilm research is still in its infancy, and clearly there is still much to learn. Even so, these little slimy blobs of cells are useful as model ecosystems. The rapid turnover allows us to see ecosystem formation and reformation on a scale that we don't often consider. The interplay of intrinsic and extrinsic factors mirrors the assembly of macroscopic ecosystems that we attempt to protect. Finally, the relationship between structure and function is a beautiful example of species interaction in a stochastic world.

The best thing about the biofilm as a model ecosystem is that it isn't an abstraction at all. Biofilm systems are a critical component of every macroscopic ecosystem on earth. When we talk about nutrient cycling, decomposition, or infectious disease we are referring to the action of biofilms. Ubiquitous in astounding numbers, these micro-ecosystems are the basis for all of the higher-order ecosystem functions.

To connect individual biofilm systems with the macroscopic ecosystem, it may be useful to think of biofilms in patchy clusters. Consider a patch of swampland, for example, that is slightly elevated from the inundated forest floor that surrounds it. It is the site of a massive tree that has long since fallen over and decomposed, leaving only this small rise where the base of the trunk and rootmass once rested. The patch has a great deal of organic matter in the soil, ample light from the gap in the canopy, and periodic flooding from the fluctuating waters. In the moist soil and on the decaying vegetation there are biofilm communities, with perhaps a wide array of constituent organisms and structures. Despite the microbial diversity, there may be some similarity of function among biofilm aggregates, for within the patch the predominant resources and stressors are somewhat uniform. For example, many of the biofilm communities may be attached to – and well-adapted to mineralize – organic substrate, limited by nitrogen availability, and subjected to periodic oxygen stress when the water levels rise. As such, we may think of this collection of biofilm communities as a sort of functional patch – a meta-ecosystem, if you like.

The connections of this patch with the greater ecosystem are obvious. This little hummock may be an important habitat for invertebrates, and it is likely a site of seed germination, decomposition and mineralization. The meta-ecosystem, like the individual biofilm communities that exist within it, is at once a complete ecosystem in itself and a component of the larger swamp. Therefore, we can ask the same questions about this hummock that we ask about the individual biofilm system. Again, questions of health and integrity hold little meaning, and so I will focus on questions of fidelity. Is the microbial community of the swamp hummock stable over time? Presented with a disturbance, will it return to its former state?

Recent advances in microbiological methods allow for an unprecedented view of the microbial community within an ecosystem. By extraction, amplification, and analysis of nucleic acids and phospholipids, the diversity and genetic structure of a particular microbial community can be characterized. One such study analyzed the stability of a soil microbial community from a wet tropical forest (Pett-Ridge and Firestone 2005). Much like our hypothetical hummock, the soils of this forest were subject to periodic saturation, and thus hypoxia is one of the main stressors of the microbial environment. In the study, forest soils were incubated under semi-saturated conditions and treated with air, nitrogen, or fluctuating air and nitrogen on short- (12 h) and long-term (4 day) intervals. After 3 weeks of treatment, microbial activity was compared among treatments and with a control field sample. In effect, this design is a suitable test for stability in the microbial community. With fluctuating stress, can the microbial community maintain its diversity and function?

The answer appears to be a qualified yes, provided that the fluctuation mimics field conditions. The 4-day fluctuation selected for a microbial community that was similar to field conditions, while the aerobic, anoxic and 12-h fluctuation produced communities that were dramatically different from field conditions. Furthermore, it was shown that the forest soil includes organisms with a wide range of survival strategies regarding oxygen availability, and considerable redundancy for each life strategy. In this case, the axiom that diversity means stability appears to be supported: the bacterial community of this soil is well-adapted to periodic oxygen stress, and thus was able to cope with the 4 day fluctuation treatment. The caveat, though, is that any substantial departure from the field stress regime – including removal of oxygen stress altogether – resulted in a significantly different microbial community. This suggests that a community's stability is relative to the selection forces acting upon it.

Resilience has also been experimentally tested in a community of soil microorganisms. In this study, sandy vineyard soil was amended with a compost of grape processing waste, and the microbial community was assessed at intervals over the next 6 months (Saison et al. 2006). The subject was the soil microbial community's capacity for response to the abrupt change in environmental condition and its ability to return to its former size, structure and activity after the disturbance. To test the effects of disturbance strength, the researchers used compost treatments at both low and high levels, in which the high level was ten times greater than the low level treatment. Upon addition of a high level of compost, the soil microbial biomass and heterotrophic activity increased dramatically within a few days and remained high for the duration of the experiment. The low compost level had a weak effect on community structure and activity after 4 days; within a few more days, the low-level community was indistinguishable from the control. The authors conclude that community resilience was observed in the low level treatment (which we may think of as a minor disturbance), and that the same community subjected to a high level (major disturbance) did not show resilience within 6 months. Perhaps the high level community would ultimately have returned to the pre-disturbance state – the experiment was not carried out long enough to tell. But a disproportionate effect is evident: the high level community, treated with ten times the compost, changed and held the change for much longer than ten times the low level effect. The take-home messages that apply to ecosystems of all shapes and sizes are (1) that community resilience is a real phenomenon, and (2) that resilience is not necessarily a function of community type; rather it is a function of the extrinsic world.

Micro- and Macro-Ecosystems

Why take this trip through the microbial world in a book about ecosystem management? The utility of such an excursion is by now, I hope, abundantly clear. Size is really irrelevant; microbial ecosystems have the same basic characteristics and dynamics as the ecosystems in which we hike, fish, and hunt. And though I have compared one to the other as though they are separate systems, they truly are not. Microecosystems are linked to macroecosystems through countless hierarchical cycles. But there are some advantages to the isolation of microbial ecosystems in the study of ecosystem assembly, structure and function. Microbial systems respond rapidly to their environment and their succession is swift. Their genomic potential is vast and highly adaptable. They can be replicated and experimented with in ways that macroecosystems cannot. In short, these little ecosystems are wonderfully instructive, and they will only become more so as new molecular techniques increase the clarity with which we see their world.

   And what can we learn from these small ecosystems that might be applicable to ecosystem management? I will highlight six lessons. (1) Both intrinsic and extrinsic factors influence ecosystem assembly and maturation. Neither can be ignored, for they are integrated: intrinsic factors are dependent upon the stochastic nature of colonization and on the stochastic constraints of the extrinsic world. And yet, even with obvious intrinsic factors in biofilms (recruitment and quorum sensing) and in macroscopic ecosystems (symbioses and feedbacks), neither can be characterized as a coherent and recurring unit. (2) Ecosystem structure and composition are directly associated with ecosystem function, but there are many structures and compositions that will allow for function. (3) Ecosystem stability may exist at certain well-defined periods, but stability is entirely dependent on the regularity of extrinsic factors like stress and disturbance. (4) Resilience likewise may occur over the course of ecosystem release and reorganization, but only if the prevailing extrinsic factors favor it. If the extrinsic factors are strong enough to select for a different structure, different complement of dominant organisms, or different function, the community will change. (5) There is nothing inherently wrong with the loss of stability or lack of resilience described in (3) and (4). (6) There is no more a normal community than there is a normal set of environmental variables.

   Let me suggest that these six points apply not only to microbial ecosystems but also to ecosystem change on a continental scale over millennia, just as they apply to the wetlands, forest, and streams that we aspire to protect today. As we set the microscope aside and return our sights to the larger world, it may be useful to keep the smallest ecosystems in mind.

References

Burmolle, M., Webb, J., Rao, D., Hansen, L., Sorensen, S., and Kjelleberg, S. 2006. Enhanced biofilm formation and increased resistance to antimicrobial agents and bacterial invasion are caused by synergistic interactions in multispecies biofilms. Applied and Environmental Microbiology 72:3916–3923.

Curtis, T., Sloan, W., and Scannell, J. 2002. Estimating prokaryotic diversity and its limits. Proceedings of the National Academy of Sciences of the United States of America 99:10494–10499.

Davey, M., and O'toole, G. 2000. Microbial biofilms: from ecology to molecular genetics. Microbiology and Molecular Biology Reviews 64:847–867

Doolittle, M., Cooney, J., and Caldwell, D. 1995. Lytic infection of *Escherichia coli* biofilms by bacteriophage T4. Canadian Journal of Microbiology 41:12–18.

Hall-Stoodley, L., Costerton, J., and Stoodley, P. 2004. Bacterial biofilms: from the natural environment to infectious diseases. Nature Reviews Microbiology 2:95–108.

Karatan, E., and Watnick, P. 2009. Signals, regulatory networks, and materials that build and break bacterial biofilms. Microbiology and Molecular Biology Reviews 73:310–347.

Matz, C., Webb, J., Schupp, P., Phang, S., Penesyan, A., Egan, S., Steinberg, P., and Kjelleberg, S. 2008. Marine biofilm bacteria evade eukaryotic predation by targeted chemical defense. PLoS ONE 3:2744.

Molin, S., Haagensen, J. A. J., Barken, K. B., and Sternberg, C. 2000. Microbial communities: aggregates of individuals or coordinated systems. In Community Structure and Co-operation in Biofilms, ed. Allison, D. G., Gilbert, P., Lappin-Scott, H. M., and Wilson M., pp. 199–214. Cambridge: Cambridge University Press.

Pett-Ridge, J., and Firestone, M. 2005. Redox fluctuation structures microbial communities in a wet tropical soil. Applied and Environmental Microbiology 71:6998–7007.

Picioreanu, C., van Loosdrecht, M. C. M., and Heijnen, J. J. 2000. Modelling and predicting biofilm structure. In Community Structure and Co-operation in Biofilms, ed. Allison, D. G., Gilbert, P., Lappin-Scott, H. M., and Wilson M., pp. 129–168. Cambridge: Cambridge University Press.

Saison, C., Degrange, V., Oliver, R., Millard, P., Commeaux, C., Montange, D., and Le Roux, X. 2006. Alteration and resilience of the soil microbial community following compost amendment: effects of compost level and compost-borne microbial community. Environmental Microbiology 8:247–257.

Stevens, T., and McKinley, J. 1995. Lithoautotrophic microbial ecosystems in deep basalt aquifers. Science 270:450–454.

Stoodley, P., Sauer, K., Davies, D. G., and Costerton, J. W. 2002. Biofilms as complex differentiated communities. Annual Review of Microbiology 56:187–209.

Ward, B. 2002. How many species of prokaryotes are there? Proceedings of the National Academy of Sciences 99:10234–10236.

Whitman, W., Coleman, D., and Wiebe, W. 1998. Prokaryotes: the unseen majority. Proceedings of the National Academy of Sciences 95:6578.

Wilson, E. O. 2002. The Future of Life. New York: Alfred A. Knopf.

Wimpenny, J. 2000. An overview of biofilms as functional communities. In Community Structure and Co-operation in Biofilms, ed. Allison, D. G., Gilbert, P., Lappin-Scott, H. M., and Wilson M., pp. 1–24. Cambridge: Cambridge University Press.

Chapter 7
Forested Ecosystems

Considering the resources that humans draw directly from ecosystems, it is easy to understand the desire for sustainability: lose the ecosystem, lose the resource. Naturally, ecosystem services are more obvious to us if they are tangible products that exist or occur at the scale of human perception. Processes that are too fast or slow, too big or small tend to be excluded from direct consideration in ecosystem-based transactions. Microbial metabolism, raindrop erosion, and ocean carbon sequestration are not commodities that we buy and sell every day. An economist might call them externalities. Nonetheless, we have increasingly become aware that the clarity of our water, the composition of our atmosphere, and the condition of our land are all critically dependent upon certain ecosystem functions. And so it makes perfect sense that much of our ecosystem conservation is based upon current and future harvest of a particular species or use of a particularly beneficial process. This is, after all, the definition of conservation – it is a philosophy of restrained use; of harvest, extraction, or exploitation in a manner that does not degrade or deplete the resource.

The image of conservation has softened a bit since Pinchot's time, though the desire to maintain a consistent harvest is still the driving force behind many efforts at sustainable ecosystem management. But there are those who consider harvest and exploitation to be contrary to the ecological qualities that are being protected. This is ecosystem protection based on cultural services, and it bears the imprint of John Muir's preservationism. In this approach it is the look, the feel, the smell of the ecosystem that should be sustained; it is characteristic species in their appropriate numbers and normal arrangement. Preservation has often been portrayed as the polar opposite of conservation, but though they have different motivating factors they arrive at the same place: perpetuation of the ecosystem in some desirable state. At the height of the Muir-Pinchot debates, the argument was indeed conservation versus preservation, and to some extent this is still disputed. After a century of ecological research, however, it is no longer the most meaningful argument. Today, we should be questioning efforts to perpetuate the ideal state for any reason in the face of abiotic and biotic forces that drive ecological change. The question should be less about which ecological state to sustain for human benefit and more about the human capacity to co-exist with inevitable ecological change.

D. J. Spieles, *Protected Land*, Springer Series on Environmental Management,
DOI 10.1007/978-1-4419-6813-5_7, © Springer Science+Business Media, LLC 2010

Nowhere have these questions – and the shifting focus of debate – been played out more than in America's forests. In this chapter I will describe several forests, each protected to provide a particular ecological service. The protection is accompanied by varying degrees of management, with results ranging from relatively unimpeded to highly restricted successional change. I use these examples to evaluate that which Muir's preservation and Pinchot's conservation both advocated – the sustainable forest ecosystem.

The Forest Sanctuary

Corkscrew Swamp Sanctuary is the largest remaining stand of old-growth bald cypress trees in the United States. It is a remnant of the vast cypress forest that covered much of southwest Florida for the last 5,000 years (Ripple 1992). Cypress wood, greatly valued for its beauty and rot resistance, was rapidly logged in the early- to mid-twentieth century, and trees that had been growing for centuries were converted into siding, pilings, shingles, pipes, and water tanks (Dennis and Maslowski 1988). Logging operations cleared much of the Big Cypress region, but it was the harvest of another commodity that first generated interest in the protection of Corkscrew. South Florida's amazing diversity of wading birds attracted plumage-hunters, who found a tidy profit in killing birds for feathers which were prized as hat adornments in elite fashion circles. Many of the desired bird species, such as the now-federally-endangered wood stork, nest in rookeries high atop trees of the cypress swamp, and those who would protect the dwindling bird numbers necessarily had to be interested in protecting some portion of the swamp from logging. The Audubon Society led the effort to purchase and protect Corkscrew in 1954, and the Swamp has since become one of the Society's most impressive sanctuaries.

Though it is only a fragment, Corkscrew Swamp is a wonderful representation of the region's pre-settlement cypress forest. The 13,000 acre sanctuary is now partially accessible by a boardwalk that winds through towering cypress, marshland, wet prairie, hardwood hammock, and pine forest habitats. The zones are distinguished by slight changes in elevation – so slight, in fact, that only 1.5 m separate the lowest and highest points in the sanctuary. Plants, microorganisms and animals respond differentially to slight topographic relief, for a few centimeters makes a great deal of difference in the face of certain ecological stressors. In the wet season, as torrential rains fill pore spaces and depressions, the great stress is oxygen scarcity. When the rains cease and the swamp dries, desiccation can be just as critical to survival. In myriad ways, the biota of the swamp is adapted to the dry and wet season. Cypress trees, for example, famously grow odd protrusions called knees from their roots, an adaptation thought to be useful in gas exchange during flooded conditions. Alligators dig wallow pits, amphibians time breeding with water abundance, invertebrates enter a dormant state during drought, microorganisms produce osmolytes to combat the hypertonic conditions, and herbaceous fugitives grow and reproduce on the floor of the swamp during the dry season. These adaptations are honed not only to the *conditions* of wet and dry, but to the *duration* of wet and dry. This is known as the swamp's hydroperiod, and it is the primary force behind the ecosystem structure and function (Duever 1978).

At this point we should be more particular about cypress varieties. The bald cypress, crown jewel of Corkscrew Swamp, is a large, long-lived tree that generally grows in rich, peaty soils. At maturity, these are ancient trees – routinely reaching an age of 900 years. Large stands of bald cypress typically occur in a topography that features some sort of periodic flooding. Lake edge swamps, river floodplains, or diffuse water flow across a gentle elevation gradient, called a *strand*, are often populated by bald cypress. A portion of Corkscrew Swamp is a bald cypress strand. The pond cypress, a variety of the same species or a close relative, depending on the authority, does not grow as large or as old. It generally grows in nutrient-poor depressions that are removed from floodplains or flowing water. In Corkscrew Swamp, pond cypress is found in the transitional areas between the bald cypress swamp and wet prairie habitats. While these two cypress varieties may be found growing together, even hybridizing, they are commonly segregated by the physical conditions of the soil, the hydrologic regime, and fire (Duryea and Hermansen 2000).

Swamps would not seem to be likely candidates for forest fire, but fire is an ecological force in south Florida. During the dry months the pine palmetto, upland hardwood forests and prairie habitats become highly flammable fuel that is frequently ignited by lightning. If conditions are dry enough, fire can spread from the vegetation to the peaty soil and burn slowly through the dry swamp. Over long periods of time, peat fires play a role in swamp formation – it is a nifty biological feedback mechanism in which the accumulation of plant-derived organic matter in the soil ultimately burns, further carving out the plants' depressional habitat. A number of species in the swamp are well adapted to, and in fact dependent upon, occasional fire. Mature bald cypress is protected from fire by virtue of its location in wet soils, but it is subject to severe damage from fire when the soil itself is burning. In Okeefenokee Swamp to the north, for example, fire has repeatedly destroyed large numbers of bald cypress trees. Some 97% of the bald cypress in a 3,000 acre Okeefenokee stand was destroyed by a peat fire in the 1950s (Cypert 1961). Peat fire can even destroy cypress in the seed bank, making regeneration less likely and enhancing the opportunity for invaders. Pond cypress can also be killed by fire, though it is thought to be more tolerant than bald cypress. Its location in soils with lower peat content reduces the likelihood of root burn. Thus, surface fire in a pond cypress stand can effectively reduce competition from fire-intolerant shrubs, conifers, and hardwoods (Ewel and Mitsch 1978).

Given the interdependence of hydroperiod, soil characteristics, and fire, the protection of a patchwork ecosystem complex like Corkscrew Swamp is a bit complicated. The hydrologic regime is delicate. A slight change in flow rate, return interval, or water quality could shift the species selection pressure and change the ecological zonation of the swamp. Long term drainage or permanent stagnant flooding could redistribute the living and nonliving features of the entire system. Changes in hydroperiod could in turn have implications for the extent, intensity, and periodicity of fire. Frequent surface fires could stabilize the cypress community, but coupled with drought they could also be detrimental. Fire suppression, on the other hand, may tip the competitive balance in favor of hardwoods and eventually change the character of the swamp (Ewel 1998). So it is with Corkscrew as it is with many other ecosystems: the critical regulators of succession can also become mechanisms of destruction.

Management of Corkscrew Swamp

Corkscrew Swamp, in the context of the Muir-Pinchot debate, is more a preservation than a conservation project. Nothing in the ecosystem is scheduled for harvest, and no fishing or hunting is allowed; the only thing designated for the use of future generations is the splendor of the swamp. There are many species that are of worthy or protection here, but the icon of this ecosystem – the flagship species of the swamp – may well be the bald cypress trees at the core of the sanctuary. But the preservation effort seeks to protect more than just these particular trees. It is the three-dimensional structure of the community that defines the sanctuary. It is the towering stand of old growth bald cypress that we seek to preserve, but it is also the transition of deepwater strand to pond cypress to wet prairie to pine key. And the preservation is all the more urgent because the vast majority of this type of system is now so rare. How then should this land be managed?

The ecosystem's sensitivity to disturbance makes the historic disturbance regime a logical starting point for management. What conditions of flood and fire drove the formation of the swamp's habitats, and can those conditions be maintained, restored, or approximated? Not every ecosystem protection effort has the benefit of historic data, but for south Florida cypress swamps like Corkscrew we have a pretty good idea. The historic hydrologic regime is a consequence of the porous limestone bedrock, minimal elevation difference across the landscape, and heavy seasonal precipitation. During the wet season, May through October, rainwater historically has infiltrated the limestone pores and flooded the swamp and prairie. The relief was so subtle that the runoff did not concentrate in rivulets, streams, and rivers, but rather moved in shallow and slow sheet flow. As rainfall tapered off in the winter and spring, the flow diminished and the water receded to the deepest depressions. In the driest years, this exposed much of the swamp floor. The historic fire regime was of course complementary, with lightning igniting the dried vegetation before the summer rains (Robertson and Fredrick 1994). Fire frequency has historically been negatively correlated with flood frequency, so that the highest elevation pinelands and prairies might have burned every 3–5 years and pond cypress communities every few decades, while bald cypress fires recur on the order of centuries (Ewel 1990).

Human encroachment and development have altered these disturbance regimes to some degree. The familiar story of humans removing water from the land brought drainage canals to within a few miles of the sanctuary. Preservationists countered the regional drainage with dikes that were constructed to retain surface water in the swamp, complete with weirs to permit the flow of excess water during particularly wet times. The dry times were also a concern, for in the drought of 1962 a fire raged for weeks in the swamp, destroying many acres of pond cypress and a few acres of bald cypress. In response, several groundwater wells were constructed so that the swamp could be irrigated when the dry season was too dry (Ingle 1974).

The present-day management plan for the Corkscrew Swamp is based on the historic disturbance regime as modified by these instances of human intervention. The hydrologic regime is an important component of the plan. Regulated impoundments and irrigation make it possible to manage the swamp's water levels to achieve the desired community. Prescribed fire is an important management tool as well, intended to prevent the encroachment of fire-intolerant species and to maintain the dominance of desirable species by reducing both fuel load and the chance of a major conflagration. This is done with prescribed burns during the winter months. Invasive plant species, including Australian melaleuca, old-world climbing fern, Brazilian pepper, and water hyacinth are controlled with physical removal, herbicides, and fire. Even some native undesirables, like *Phragmites* and coastal plain willow are managed to prevent excessive encroachment. In these ways the conditions of the remnant ecosystem and its wildlife habitat are controlled in the pristine state for the benefit of the human visitor (Audubon of Florida 2009).

Corkscrew Swamp is absolutely worthy of protection, but it is very much a contrived system. Water pumps, dikes, prescribed fire, herbicides – the sanctuary is even enclosed by a fence. It is an ecosystem on a respirator. The argument in favor of such complete management is simple: this is the last remaining example of a pristine bald cypress swamp; for its beauty, for its importance to wildlife, and for its place in our nation's natural legacy it must be preserved. I have no doubt that, with enough human intervention, the swamp can be preserved in the pristine state indefinitely. But the ultimate reality is that ecosystems change, and by choosing to allay one sort of change we are inviting another.

Let me suggest three ways in which the management schemes at Corkscrew Swamp may lead to unintended consequences. First, replacement of the natural fire regime with prescribed burning has changed the seasonality and character of fire in the swamp (Kirkman et al. 2000). Prescribed fires during the wetter winter months are likely to burn over a smaller area, for a shorter duration, and with a lesser intensity than natural fires. That is the point, of course, for intense natural fires could be more destructive to the cypress trees, and prescribed burns reduce the chance of destruction. The practice of irrigating during drought also helps to prevent major fires. But the longer the swamp goes without a major fire, the more the peat will build; so while fire danger is reduced in the short term, it may increase in the long term. When the big fire finally comes, it could be intense. Even if irrigation can prevent a major fire indefinitely, it could lead to a second consequence: topographic change. As organic detritus builds in the swamp, it changes the landscape morphology. Given a long enough time without a major burn or aerobic decomposition, slight depressions become slight mounds. Drier, higher patches of land may be points of colonization for a different sort of plant community. Finally, the effort to maintain a particular biota in a specific arrangement by eliminating extremes of the hydroperiod and fire regime could select for a functionally narrow community. Limitation of the range and type of disturbance events will eventually limit the response diversity of the ecosystem.

Corkscrew Swamp is representative of forested systems that are protected for the cultural services they provide. Our impulse to preserve is contrary to the nature of directionless and stochastic succession. Preservation can be achieved with human intervention, and while it is not without long-term consequences it can be maintained, perhaps for years or decades. But there is a great contradiction here – while such efforts seek to sustain the ecosystem in its ideal state, the prevention of change is itself unsustainable.

The Monongahela Experiment

Corkscrew Swamp is, in several ways, an anomaly; very little of the protected forest in the United States has the same status or level of management. Most of it is more correctly described as conservation land, as it is intended for some collection of uses. How, then, do we reconcile the shifting patchiness of succession with the ecosystem ideal in our multiple-use forests?

In West Virginia there is a nice example of the range of national forest management schemes. The Monongahela National Forest covers about 920,000 acres in the northeastern part of the state. This is no virgin forest – it has been logged repeatedly, and most of the trees are less than a century old. There are only a few scattered patches of old growth that somehow managed to evade the saw. Established as a national forest in 1915, it now supports a number of different habitats that are managed to provide a wide variety of human services. In addition to commercially valuable timber, the forest is the site of natural gas wells, camping and hiking, fishing and hunting, and livestock grazing. Management techniques include timber harvest "used to emulate naturally occurring disturbances" like fire and windstorms, or to create and maintain certain tree stands. Wildfire is suppressed to protect "forest resources and investments, as well as nearby private property," and it is replaced by prescribed fire as appropriate. Tree planting accompanies harvest to stimulate regeneration (USFS 2006).

The uses and management tools are numerous, but not every portion of the forest is treated in all these ways. In fact, 115,000 acres are devoted to eight wilderness areas, where only low-impact recreational activities are permitted and management is limited. One of these, the Otter Creek Wilderness Area, is adjacent to an area of silviculture research called the Fernow Experimental Forest. In the Fernow Forest a number of harvest techniques, including even- and uneven-aged management, patch cutting, and selective cutting, are compared with an unharvested control area to determine the effects of each on log quality, erosion, and watershed characteristics. The juxtaposition of Fernow and Otter Creek, developing as they have under similar geomorphic and climatic conditions but with different management schemes, provides an interesting glimpse of alternate successional pathways.

In effect, these are the makings for a semi-natural experiment. The variety of management regimes – including, in the case of the wilderness area, no management at all – may be compared with the old growth and control treatments to shed some light on forest succession on protected land. In particular, we might consider questions of forest stability, integrity, and health. The holistic viewpoint might lead us to hypothesize that while different management schemes yield different types of forest community, the unmanaged wilderness will revert to its domain of attraction and come to resemble the old growth in terms species composition and function. The best part about this experiment is its age – the forest management treatments at Fernow have been in place for 50 years, and the wilderness area has been unmanaged for nearly a century.

But we immediately encounter a problem. Our experiment pre-supposes that the old growth represents a single ideal state which we may use as a basis for comparison, and this is simply not the case. The Monongahela Forest management plan (USFS 2006) states this plainly:

> Old-growth forests can display a wide variety of vegetative conditions, depending on factors such as species composition, stand age, environmental conditions (climate, geology, topographic position), and soil productivity. The appearance and function of old growth differs dramatically depending on forest type (e.g., spruce-fir vs. oak-pine vs. mixed mesophytic). Some forest types do not support much plant or wildlife diversity no matter how old they grow. Others can be species rich at a fairly young age and continue to add diversity and complexity as they grow older.

As it is, some of the Monongahela's old growth is dominated by white pine; others are spruce-hemlock, spruce-hardwood, or hemlock dominated. Furthermore, none of these are stand-alone units, disassociated from the rest of the forest. Each is comprised of species that are interspersed differentially with adjacent communities. The various patches of old growth have different understory compositions, different herbaceous communities, different amounts and types of standing dead trees, and different woody debris. This should come as no surprise. These small patches of old growth each have their own peculiarities of topography, aspect, and conformation. This, of course, means that they have different soil types and moisture regimes. They have different histories of exposure to fire and grazing. They have been affected in different ways by chestnut blight, hemlock wooly adelgid, beech bark disease, and gypsy moth. Each has its own set of historic contingencies, and each continues to change over time. We have no reason to expect them to be alike.

The rest of our experiment follows suit. Each of the managed stands at Fernow, as you might expect, have yielded communities of different composition and diversity (Schuler 2004). And while all are unique, none are stable. All of the treatments are experiencing the loss of beech, oak and cherry, ostensibly due to fire suppression, disease, and grazing pressure (Shumway et al. 2001). Even in the managed stands, the future species composition is expected to transition from a beech-maple-oak-hickory complex to a maple-basswood dominated community (Schuler 2004). Despite a half-century of management, these forest ecosystems are not in equilibrium.

How about the unmanaged Otter Creek wilderness? After a century of succession, has the wild forest come to resemble old growth conditions? The answer is yes, and no. In a general sense, the forest here is typical of a mixed mesophytic Appalachian forest. Low elevation communities support maples, birches, beech, black cherry and tulip tree; hemlocks and hardwoods dominate intermediate elevations, and spruce is dominant at high elevations. The forest floor is moist and slightly acidic, supporting ferns, liverworts and mosses, and fungi. Some of these environments share some characteristics with the remnant patches of old growth.

But the most striking ecological feature in the wilderness area is the dominant shrub at all elevations, the great rhododendron. This native, shade tolerant shrub grows in impenetrable thickets 20–30 ft in height throughout the Otter Creek Wilderness. It is an aggressive competitive dominant that is difficult to remove once established. Without a doubt, the dense rhododendron inhibits the growth of other plant species and limits habitat suitability for some wildlife species. It has been described as a "serious woody weed" in the southeastern United States. Its present dominance is likely the result of the great opportunity for colonization presented by historic logging, the acidification of soils by acid deposition, and fire suppression. As an invader, it appears to be a "driver" and not just a "passenger." It reduces forest floor light levels and significantly decreases the plant species richness of areas it invades. Its litter is nutrient poor and slow to decompose, and so it depresses microarthropod diversity and earthworm activity. The resulting effects on soil composition give rhododendron seedlings an advantage over other plant species of the understory. Though native, it is likely more pervasive than it ever was in the pre-settlement Appalachian forests. As a result, the unmanaged wilderness is different than it was prior to logging.

What can be learned from the Monongahela experiment? For starters, it seems clear that there is no single place in these million acres that can be identified as the quintessential Appalachian forest. There simply is no single ecosystem unit here; species are distributed along a multitude of gradients and according to idiosyncrasies of place and time. It must follow, then, that a definition of ecosystem integrity and health, which both require a reference system, cannot easily be justified. Second, there is no timeframe after which a portion of this forest will reach some stable state and cease to change. In part, this lack of stability is due to human-induced pressure, like acid deposition and introduced disease. But it is also due to changes in soil characteristics and mechanisms of dispersal that are caused by natural processes and the resident biota. Finally, we can conclude that the Otter Creek forest is apparently not resilient, for a century after logging it still has not returned to its pre-logging state. But this is not due to the lack of some ethereal quality in the pre-logging forest community. Rather, there is evidence that the major change – rhododendron encroachment – is a consequence of changing selection pressure and opportunistic response.

The Monongahela National Forest is unlike Corkscrew Swamp, but it is no less beautiful. As with Corkscrew, some old growth areas of the Monongahela have been declared sanctuaries – in this case, they are called National Natural Landmarks. According to the Monongahela National Forest Management Plan, the goal in these areas is to "maintain virgin forest characteristics." If the Monongahela experiment can teach us anything, it is that this vision is indefinable and unattainable. Nevertheless, a particular forest can be managed to achieve and maintain an ideal state, and much of the American approach to forest management is intended to do just that. But if the management is suppressing succession, isn't the stand more a museum than an ecosystem?

Wilderness Management

As I write this, the United States is celebrating the 45th anniversary of the idea that made the Otter Creek Wilderness possible. On September 3, 1964, President Lyndon Johnson signed into law the Wilderness Act, "to establish a National Wilderness Preservation System for the permanent good of the whole people." Since its inception, 109 million acres have been protected from most high-impact human encroachment. In a few cases mineral extraction is permitted on legitimate claims, though new claims are prohibited. Some livestock grazing occurs on wilderness lands, and hunting, trapping, fishing, and non-motorized recreation are allowed. For the most part, that's it: no hotels, no road building, no logging, and no dam construction. The idea is to allow these areas to exist "without permanent improvements or human habitation … primarily affected by the forces of nature."

As close as this law comes to allowing land to exist in an unmanaged state, it doesn't quite dispense with the holistic ideal of the ecosystem. The Act includes phrases like "preservation of wilderness character," preservation of "natural conditions," and "land retaining its primeval character" which all hint that these lands are to be maintained in a particular ecological state. Accordingly, wilderness areas are not quite unmanaged. The law specifically allows for the control of fire, insects, and diseases, as in such cases where nearby private property or some common good is threatened. But this brings up an interesting question. Is it a particular successional state that makes an area worthy of wilderness designation? What if the "forces of nature" conspire to dramatically change the ecological character of the system being preserved? To what extent should wilderness be managed?

This very scenario is being played out in northern Minnesota. On July 4, 1999, a windstorm of incredible strength hit the Boundary Waters Canoe Area Wilderness of the Superior National Forest (Mlot 2003). The storm was brief but destructive; reaching straight-line speeds of 100 miles/h, it knocked down trees over about a third of the wilderness area's 1.1 million acres. The damage was most intense over a 90,000 acre swath in which the entire forest canopy was destroyed. This was a disturbance of major proportions – the likes of which might happen once every 1,000 years or so – and it immediately changed the character of the wilderness area. Not only were there suddenly 100 t of woody debris per acre in the most severely affected portions of the forest, there was also increased light penetration in formerly shaded areas, an abrupt change in wildlife distribution and behavior, and abundant exposed soil that was ripe for colonization. Above all, there was now a great risk for colossal wildfire in an area that hadn't experienced a major fire since 1910.

All of this threatened to change the "primeval character" of this wilderness and tested our national resolve to leave wilderness areas to the "forces of nature." After all, if humans weren't here, that really big windstorm would still have occurred, and it probably would be followed by a really big fire.

Humans, it was decided, must intervene, for that really big fire could be devastating. The Forest Service estimated that the storm had increased the fuel load ten to twenty fold in some places, that the potential rate of fire spread had tripled, and that the potential fire intensity had doubled over pre-windstorm conditions. It was determined that this could cause ecological harm, for unchecked fire could be greater in scale and intensity than fire's historical range of variation. It would also likely threaten the recreation potential of the area, as well as public safety and private property if the fire should escape the wilderness. After the case for intervention was made (in an eight-pound environmental impact statement), it was determined that the fuel load must be reduced (Mlot 2003). The result has been a monumental fuel reduction effort, mostly by prescribed burns on 75,000 acres of forest.

This was the first major test of the hands-off principles of the Wilderness Act. Confronted with major ecological change, the impulse for management proved to be irresistible. Granted, the issues of public safety, recreational restrictions and damage to private property are indisputable. There are even some convincing ecological arguments for intervention. But some of the reasoning in favor of fuel reduction stems from a desire to keep the Boundary Waters looking as we think it should look and acting as we think it should act. For example, the Superior National Forest Management Plan (USFS 2004) includes terms like "apparent naturalness" and "scenery management" for the purpose of maintaining "scenic integrity." Scenic integrity is defined as "the degree to which a landscape is usually perceived to be 'complete'." A more holistic ideal has never been set to words. Apparent naturalness and scenic integrity? Frederick Law Olmsted would be proud.

And so, by way of fuel reduction and prescribed fire, the big wind may not be followed by the big burn after all. But what of the forest? Is it recovering from the storm, and is the recovery on trajectory to return the ecosystem to its former, pre-windstorm state? Of course it is recovering. Species formerly inhibited by shade, including balsam fir, black spruce, northern white cedar, and dogwood, have experienced explosive growth. Sun-loving species dispersed by wind and birds, like fireweed, currant, and blackberry have also thrived. Some species that are typically seen later in succession, such as white cedar and paper birch, survived the storm reasonably well and are now intermingled with the colonizing r-strategists. The vast majority of post-storm growth is native, but its distribution and composition have changed. Other native species were hit hard by the wind; jack pine, aspen, and much of the old-growth red and white pine trees were taken out. It might be difficult for these species to be a significant part of the post-storm forest, for climate change has altered the competitive playing field. Hardwood species like red maple and oak are advancing northward and beginning to replace the conifers that had dominated the area for centuries. Prescribed fire may keep the hardwoods at bay for a time, but climate change is not so easily managed. So while the ecosystem is indeed in recovery, it will likely not ever be the same; in the words of Christine Mlot (2003), "the forest of June 1999, much less of 1899, won't be back any century soon." We could blame this on a century of fire suppression; we could blame it on the freakish intensity of the windstorm; we could blame it on climate change. But why should we blame it on anything? Is the forest less of an ecosystem now than it was prior to the storm?

Healthy Forests

Humans had little say in the matter of the Boundary Waters windstorm, but we have a way of manipulating or minimizing disturbances when we can. Fire in particular is an ecological force that by presence or absence has shaped many of the ecosystems we see today. Humans, of course, have a long history of using or preventing fire as a way of managing American ecosystems. Both Native Americans and early European American settlers used fire to clear land, to drive game, to signal friends and to wage war on foes. Native Americans commonly used fire to manage ecosystems – clearing land for agriculture, removing forest undergrowth, and creating early successional edge habitat as a prime hunting environment. It was this highly manipulated environment, released from management as Native Americans succumbed to disease and displacement, that European Americans perceived as the North American wilderness. As Stephen Pyne (1997) has noted, "the virgin forest was not encountered in the sixteenth and seventeenth centuries; it was invented in the late eighteenth and early nineteenth centuries."

As American society became more sedentary, fire continued to be used as a land management tool, but wildfire became a real and perceived threat to property. In the twentieth century fire suppression took on a greater importance, and driven by the timber industry's marketing blitz it became a de facto national policy (Pyne 1997). At about the same time, ecologists were beginning to consider fire as a natural event with biological implications and suggesting, as described in the Leopold Report (1963), that long-term fire suppression was not the best forest management strategy. Rather, the Leopold Report advocated the use of fire to guide the ecosystem to and maintain it in a desirable state. E.P. Odum, author of the most influential ecology text of his day, concurred: "fire is and has been in the past an important factor in many environments and … it can be used as a tool in management on a much wider scale than is generally realized" (Odum 1953). The central concept here is largely Clementsian: that fire of low intensity may be used to arrest an ecosystem at a particularly desirable successional stage and to prevent a destructive conflagration. This launched an era of fire research in the 1960s and 1970s and marks the renewal of an ancient Native American practice: prescribed burning.

Prescribed burning treats fire as a tool with which an ecosystem may be manipulated. Along with the popularity of prescribed fire has come the heightened sense that the managed state is the healthy state. A forest with excessive fuel accumulation, and by extension the intense and unpredictable fire that it might generate, are by definition unhealthy. This attitude became national law in 2003 with the passage of the Healthy Forests Restoration Act, better known as the Healthy Forests Initiative. The focus of the law, like the management priority in the Boundary Waters, is fuel reduction on federal lands. "Fuel reduction" means biomass removal by harvest and prescribed fire. The idea is that fuel reduction will protect human interests from the risks of catastrophic wildfires, but the law suggests that there are ecological benefits as well. It will, in fact, "protect, restore, and enhance forest ecosystem components" by promoting the recovery of threatened and endangered species, by improving biological diversity, and by enhancing productivity and carbon sequestration. The law not only protects human communities and commodities, but also "maximizes the retention of large trees, as appropriate for the forest type, to the extent that the trees promote fire-resilient stands." And so, by law, the post-storm Boundary Waters Canoe Area Wilderness *is* in need of improvement. It is our national policy that this forest should be guided back to its appropriate state. Clementsianism is alive and well.

Unmanaged Forest Succession

We have considered forest succession in the highly managed Corkscrew Swamp, in the Otter Creek Wilderness that was once managed but then left to succession, and in the Boundary Waters Wilderness that was left to succession but then managed anyway. In closing we should consider an old growth forest that encountered a major disturbance and was then left to succession that is (so far) relatively unimpeded by human management. Such a scenario has occurred in southwest Washington.

On May 18, 1980, Mount St. Helens erupted violently, knocking down thousands of acres of trees and spewing ash over a wide area. Some 24 megatons of thermal energy was released, causing the air to reach 660°F in some places (Dale et al. 2005). Prior to the eruption, a portion of the blast area was managed under the auspices of the Gifford Pinchot National Forest while the rest was in state and private ownership. Much of the land had been the site of mineral and timber extraction, though some forest north of the volcano was old growth and primarily used for recreation (Dale et al. 2005). This area was devastated by the blast and buried with ash to a depth of ten inches up to ten miles from the crater. Could an ecosystem ever recover from such a disturbance? We know that it can, of course, because Mount St. Helens has erupted before. Every few centuries a portion of the forest has been burned, buried, or cooked by ashfall, mudflows and hot gasses emitted by the volcano (Dale et al. 2005). Each time, living things have re-colonized the disturbed land. Holistic ecologist Ernest Partridge (2000), upon surveying this most recent devastation, predicted that the forest ecosystem "will once again become what it was before: a northern conifer rainforest." This, according to Partridge, "is what ecologists correctly call a 'climax stage.'"

The reason for Partridge's confidence is his view of the surrounding landscape. At areas "up and down the Cascade Range" similarly devastated by historic eruptions, there are "various stages of succession and recovery." These are progressing toward "an identifiable 'type' of integrated life community" (Partridge 2000). The surrounding peaks do indeed support vegetation at various stages of succession, and the pre-eruption

Mount St. Helens had its own unique set of communities. The landscape around the mountain was largely forested, with a mix of conifers and hardwood species, all with individual ranges varying by elevation, physicochemical factors, and the time since the last disturbance. There were also "meadows, wetlands, cliffs, seeps, and avalanche paths," all with their own unique assemblages (Dale et al. 2005). In fact, the biotic community was as variable as the landscape itself. The "type" ecosystem that existed before the eruption was really a host of intermingled types; any reference "type" would need to be specifically defined in terms of soil, altitude, slope and aspect, nutrient availability, and other critical factors. Furthermore, given that species are distributed independently, the "type" ecosystem would need to be further sub-divided by species complement. The idea of the "type climax" is acceptable only if we allow that different combinations of abiotic factors and colonization opportunities result in different climaxes. As a result, we would have to acknowledge thousands of climax states in the Cascade Range, which in my mind is the same as saying there is no climax community at all.

In any case, succession has progressed in the quarter century after the Mount St. Helens eruption, just as it has after each eruption for thousands of years. But the communities on Mount St. Helens are not progressing "by repeatable deterministic mechanisms" toward some climax stage (del Moral et al. 2005). Community assembly is instead highly dependent on site characteristics and on the chance arrival and colonization of various species. The substrate is critical: are prospective colonizers able to exist on fine pumice, mudflow residue, tephra, or coarse pumice? Is the site on a ridge or plain? Are the nutrient and moisture requirements of the prospective colonizer met? Is the species tolerant of the stress regime? What sorts of relic vegetation survived the blast, and how close is it to the point of colonization? All of these conditions and many more have driven the development of great community heterogeneity in the blast zone, even in areas that appear to have similar characteristics. In short, the first quarter century of succession after the eruption has been rather chaotic, as summarized by Roger del Moral and others (2005): "Apparently, several alternative, equally 'natural' communities can develop after an intense disturbance, and the one that ultimately results is initially poorly predictable. Mature communities will retain a residual of unexplainable variation due to historic accidents (contingencies), stochastic invasion patterns, and landscape effects." After another century of succession, will Partridge's "type climax" exist anywhere in the Mount St. Helens blast zone? It might. But why restrict succession to such a narrow set of possibilities?

And so, we return to the question of ecosystem sustainability. In this chapter I have highlighted examples of forests that we protect for the services they provide. In each example, we have seen the effects of disturbances recent or long past. The biotic communities of each forest are a function of the particular disturbance regime, the specific abiotic conditions, and the chance opportunities for colonization. In each case we see the remarkable ability of ecosystems to recover from disturbance, but we also see that the pre-and post-disturbance ecosystem can differ substantially. What, then, is a sustainable ecosystem? What are we trying to sustain? In our protected forests it seems that we commonly endeavor to sustain a particular set of species which we see as the ecological legacy for that place – much like the utilitarian conservationist seeks to sustain species that are valuable for harvest. There is nothing wrong with the desire to protect certain species, but we must recognize that each community is made possible by a multitude of variables and contingencies. With great effort, we can to some extent manage variables to achieve and maintain a particular community, but this is hardly sustainable in the long term. Instead, perhaps our efforts would be better spent on the protection of an ecosystem's capacity for response.

References

Audubon of Florida. 2009. Natural Resources Management Program, Corkscrew Swamp Sanctuary, Naples, FL.

Cypert, E. 1961. The effects of fires in the Okefenokee Swamp in 1954 and 1955. American Midland Naturalist, 66:485–503.

Dale, V. H., Swanson, F. J., and Crisafulli, C. M., eds. 2005. Ecological Responses to the 1980 Eruption of Mount St. Helens. New York: Springer.

del Moral, R., Wood, D. M., and Titus, J.H. 2005. Proximity, microsites, and biotic interactions during early succession. In Ecological Responses to the 1980 Eruption of Mount St. Helens, ed. Dale, V. H., Swanson, F. J., and Crisafulli, C. M., pp. 93–109. New York: Springer.

Dennis, J. V., and Maslowski, S. 1988. The Great Cypress Swamps. Baton Rouge: Louisiana State University Press.

Duever, L. 1978. Dry season, wet season. Audubon 80:120–130.

Duryea, M., and Hermansen, L. 2000. Cypress: Florida's Majestic and Beneficial Wetlands Tree. Circular 1186. Gainesville: School of Forest Resources and Conservation, Florida Cooperative Extension Service, Institute of Food and Agricultural Sciences, University of Florida.

Ewel, K. 1990. Swamps. In Ecosystems of Florida, ed. Myers, R., and Ewel, J. pp. 281–323. Orlando: University of Central Florida Press.

Ewel, K. 1998. Pondcypress swamps. In Southern Forested Wetlands: Ecology and Management, ed. Messina, M., and Conner, W., pp. 405–420. Boca Raton: CRC Press.

Ewel, K., and Mitsch, W. J. 1978. The effect of fire on species composition in cypress dome ecosystems. Florida Scientist 41:2–31.

Ingle, B. 1974. Corkscrew Sanctuary: use of the market for preservation. Boston College Environmental Affairs Law Review 3:647–686.

Kirkman, L., Goebel, P., West, L., Drew, M., and Palik, B. 2000. Depressional wetland vegetation types: a question of plant community development. Wetlands 20:373–385.

Leopold, A. S., Cain, S., Cottam, C., Gabrielson, I., and Kimball, T. 1963. Wildlife management in the national parks. American Forests 69:32–63.

Mlot, C. 2003. The perfect windstorm study. BioScience 53:624–629.

Odum, E. P. 1953. Fundamentals of Ecology. Philadelphia: W. B. Saunders

Partridge, E. 2000. Reconstructing ecology. In Ecological Integrity: Integrating Environment, Conservation, and Health, ed. Pimentel, D., Westra, L., and Noss, R. F., pp. 79–98. Washington: Island Press.

Pyne, S. J. 1997. Fire in America: A Cultural History of Wildland and Rural Fire. Seattle: University of Washington Press.

Ripple, J. 1992. Big Cypress Swamp and the Ten Thousand Islands: Eastern America's Last Great Wilderness. Columbia: University of South Carolina Press.

Robertson, W. B., and Fredrick, P. C. 1994. The faunal chapters: contexts synthesis and departures. In Everglades: The Ecosystem and Its Restoration, ed. Davis, S., Ogden, J., and Park, W., pp. 709–740. Boca Raton: St. Lucie Press.

Schuler, T. 2004. Fifty years of partial harvesting in a mixed mesophytic forest: composition and productivity. Canadian Journal of Forest Research 34:985–997.

Shumway, D., Abrams, M., and Ruffner, C. 2001. A 400-year history of fire and oak recruitment in an old-growth oak forest in western Maryland, USA. Canadian Journal of Forest Research 31:1437–1443.

USFS. 2004. Superior National Forest Land and Resource Management Plan. Washington: United States Forest Service, United States Department of Agriculture.

USFS. 2006. Monongahela National Forest Land and Resource Management Plan. Washington: United States Forest Service, United States Department of Agriculture.

Chapter 8
Grassland Ecosystems

The ecosystems that we protect were forged by a long history of disturbance, stress, and response. In some ecosystems the disturbance and stress regimes have been harsher than others. Frequent, intense disruption can make a place nonviable for many species, as can severe and prolonged physiological limitations. As a consequence, we see simpler ecosystems in conditions of extreme disturbance and stress. The organisms that inhabit such ecosystems manage to exist as they do because they are in some way tolerant to the stress, adapted to the disturbance regime, and/or highly specialized to exploit some particular resource. The community- and system-level features of such environments consequently include some unique and fragile associations. When the disturbance regime changes abruptly, or when new stress is imposed, the former associations can rapidly dissolve, to be replaced by a new community as time and conditions allow.

American grassland ecosystems are the product of stress and disturbance, just as forests, wetlands, and lakes are. But our prairies hold a special place in American environmental history and in our current land protection effort, for these systems that once covered so much of the nation are now so few in number and so small in area. Some remnants still exist under something like the disturbance and stress regimes of their recent history, and some are protected or undergoing restoration. Protection and restoration often mean management, and so a wide range of modifications and interventions are used to minimize the departure from the conditions of legacy. But stress and disturbance regimes have a way of changing over time. Maintaining these highly adapted ecosystems amidst changing rules of adaptation is challenging to say the least.

In this chapter I will explore the stress, disturbance, and response of some American grassland ecosystems. In each case the conditions of disturbance and stress drive structure, function, and human management schemes.

Grasslands, Short and Tall

Grasslands once covered approximately 162 million acres of the Great Plains, stretching from Alberta to Texas and Ohio to the Rocky Mountains. Prairie regions, as differentiated by water availability and dominant species, have all been devastated

D. J. Spieles, *Protected Land*, Springer Series on Environmental Management,
DOI 10.1007/978-1-4419-6813-5_8, © Springer Science+Business Media, LLC 2010

by agriculture. Of the wettest environments, the tallgrass prairies, only about 4–13% remains; mixed grass and the driest shortgrass prairies have fared only slightly better. But only a few *hundredths* of a percent of the original grassland extent is protected from future development, making these remnants some of the most threatened ecosystems in the North American conservation effort (Samson and Knopf 1994; Samson et al. 2004).

To characterize the management of prairie remnants, I will first compare dry shortgrass ecosystems with a tallgrass system that really isn't very dry at all. The wetter of these two examples is the Curtis Prairie in Wisconsin. Its fame is not due to its size, for it covers only about 60 acres. Nor is it some ancient remnant, like a Corkscrew Swamp of the prairie world. On the contrary, whatever prairie was here in the early nineteenth century was plowed and farmed for 100 years. But Curtis Prairie is a special ecosystem, for it is the world's oldest restored grassland. Initially overseen by Aldo Leopold, among others, Curtis Prairie was planted with native seed and remnant prairie sod in the 1930s. At about the same time, some 4 million acres that would become the National Grasslands of the western Great Plains were being reclaimed from agriculture after ill-fated farming attempts of the dust bowl. These western National Grasslands are primarily short and mixed-grass prairies that are and have historically been multiple-use resources, intended to meet the needs of the rancher, the outdoor enthusiast, and prairie wildlife.

These two prairies are different in many respects. The size is an important difference, as are the pressures of the surrounding landscape. The land use demands and expectations are dissimilar, for one ecosystem is a demonstration while the other is simultaneously wilder and more suitable for provisional services. Most importantly, these prairies experience different ecological conditions. One obvious disparity is in average annual precipitation: Curtis Prairie gets over 36 in./year, while only about 12 in. falls on the prairies of the western Great Plains. This translates into different stress regimes. But these two types of grassland also have some things in common. They are both fragments of the once-vast area of North American grassland, and both were reclaimed in the same era. Both are managed ecosystems; in each case managers use tools like prescribed fire and grazing to control undesirable species and encourage desirable ones. And, as we shall see, the goal in both systems is the maintenance of a specific community in spite of succession.

Curtis Prairie

Curtis Prairie is part of the University of Wisconsin's Arboretum in Madison. At the dedication of the arboretum in 1934, Aldo Leopold summarized its purpose: "Our idea, in a nutshell, is to reconstruct, primarily for the use of the university, a sample of original Wisconsin – a sample of what Dane county looked like when our ancestors arrived here in the 1840s" (Leopold 1988). Other have pointed out that the grasslands of the American Great Plains, and especially the fingers of prairie that extended north, east and south were almost entirely created by human activity, "the product of deliberate, routine firing" (Pyne 1997). Nonetheless, the condition of the ancestral land was perceived as an ecological legacy, and the Curtis Prairie experiment was born. The endeavor was a novel one at its inception, and its early research on methods of ecosystem re-creation has made the Arboretum a cradle of restoration ecology. It was here that different planting methods were first assessed and that the use of fire as a prairie management tool was firmly established (Anderson 1973). The location of Curtis Prairie places it on the historic border between the tallgrass prairie and hardwood forest ecosystems. While there are regional prairie remnants yet in existence, most areas of south-central Wisconsin, if left to the devices of succession today, tend toward wooded ecosystems. Thus, beating back the forest is a priority; a good deal of effort has been required to establish and maintain the oldest restored prairie in the world.

The humble beginnings of Curtis Prairie featured an agricultural field near a remnant of unplowed prairie. After about a century of cultivation and use as pasture, agricultural activity ceased in 1932 and succession ensued on the abandoned field (Cottam and Wilson 1966). The first colonizing plants were agricultural weeds, for their seeds existed in the seedbank and dispersed to the site from adjacent fields. This was no primary succession from sterilized volcanic or glaciated substrate. Rather it was secondary succession, and the conditions of the previous land use were apparent in the early stages of succession. Within a few seasons the field was dominated by weeds: quack grass, Kentucky bluegrass, and Canada bluegrass (Curtis and Partch 1948). There was an interest in restoring on the land an ecosystem similar to the oak savanna grasslands that had existed prior to European American settlement, and the throes of the Great Depression meant that labor was readily available from the nearby Civilian Conservation Camp. In 1936 and 1937 native sod and nursery-reared prairie forbs were transplanted into the field, but after a few growing seasons it quickly became clear that the transplants were being out-completed by weeds. Thus began a systematic study of the effects of fire on grassland plant competition. Plots were designed to test the effects of annual and biennial fire administered at different times of the year against adjacent control plots. Then the plots were burned on schedule from 1941 to 1946. The result was a drastic reduction in bluegrass cover and a subsequent invasion of annual species in burned plots. The effect of fire on introduced prairie species was variable: some were harmed, some unaffected, and some responded vigorously. Overall, the conclusion was that fire can be a useful management technique "in certain regions where prairie is not the climatically favored formation" (Curtis and Partch 1948).

Prairie plant introductions continued in the 1950s, "some of them carefully planned, some haphazard" (Cottam and Wilson 1966). So, in addition to fire regime, there were differences in planting diversity and intensity among plots. By the early 1960s the introduced plants had begun to sort themselves out according to a third variable: moisture regime. Some plants, it seems, favored a drier soil than others, some thrived in a mesic moisture regime, and still others found competitive advantage in wetter areas of the prairie. The boundaries of moisture regime and plant occurrence were neither sharp nor mutually exclusive, but specific community zones were clearly becoming apparent.

And so, after 30 years of growth and development at Curtis Prairie, the logical question – is this restored prairie similar to native prairie remnants? – was ambiguous. Which zone of Curtis Prairie should be compared with which native remnant? It would only make sense to compare areas with analogous fire history and moisture regime, though even those could only be compared with the caveat of dissimilar colonization. Furthermore, the native prairie patches were not static entities as the restored prairie progressed through development. The drought of the 1930s had subjected native remnants to shifts in community organization, and these were followed by site-specific recoveries and reorganizations when the rains returned. To which manifestation of a particular native prairie patch should the restoration be compared? There simply was no "typical" native Wisconsin prairie for comparison. The best proxy for a site-to-typical-site comparison was, and still is, an aggregate characterization of the plant community by frequency and abundance. In this respect, despite the ambiguity, the Curtis Prairie was by 1961 supporting prairie plants in assemblages that resembled those of native prairies, though the restoration still harbored many more weedy and non-prairie species than native sites. The conclusion at this time was a mixture of individualistic realism and holistic optimism: "it will take many years for the Curtis Prairie to become truly representative of native prairies" (Cottam and Wilson 1966).

By the new millennium, some 65 years after the restoration was initiated, Curtis Prairie still had not become similar to the reference remnant in terms of species abundance, distribution, or dominance (Kucharik et al. 2006). Some characteristics, like species richness and productivity were similar when compared in areas of similar fire regime. Functionally, the two ecosystems differed in terms of soil respiration and carbon sequestration, though both communities responded similarly to seasonal variability and were alike in aboveground net primary production. Other functions have not yet, and may never, converge. Thus it seems that the restored prairie had not achieved "typical prairie" status even after 65 years.

Today, intensive management to achieve and maintain the ideal continues. Recent storm water flooding has brought an influx of sediment, seed, and nutrients – suitable conditions for the invasion of an undesirable plant called reed canary grass. Woody species continue to encroach, including aspen trees that are not easily controlled with fire. Reed canary grass and aspen trees – both technically native plants themselves – are among the threats to the native plants of the restored prairie. Thus the current management scheme includes hydrologic modification, herbicide application, prescribed fire, girdling, and planting of desirable species. In the face of changing stress and disturbance regimes, Aldo Leopold's vision is being maintained only with considerable effort.

The National Grasslands

America's National Grasslands are now part of its ecosystem protection effort, but they did not come about through some visionary effort to preserve a pristine ecological environment. On the contrary, they were formed in response to a national tragedy. Post-Civil War settlers of the Great Plains began what would ultimately be recognized as an ecologically abusive lifestyle, characterized primarily by overgrazing and cultivation without erosion control. The onset of severe and prolonged drought in the 1930s ushered in the dust bowl and subsequent New Deal programs aimed at soil stabilization and landuse reform (Hurt 1985). One such program was the creation of national grasslands (though they were not so named until 1960) by federal acquisition of degraded land in private ownership. Much of the land that would be acquired through this program was shortgrass prairie, and since it was too dry to be even marginally arable, much of it had never been plowed (Weaver et al. 1996).

Ironically, this land had once been federally owned but was distributed to private ownership through the Homestead Act of 1862. Now returned to governmental control, the land utilization initiative of the 1930s was intended "to develop a program of land conservation and land utilization, including the retirement of lands which are submarginal or not primarily suitable for cultivation in order thereby to correct maladjustments in land use" (Olson 1997). Under the auspices of the Soil Conservation Service (1938–1954) the condition of the acquired land was improved with erosion control measures, including revegetation. Seeding had the initial purpose of stabilizing the soil, but the ultimate purpose was to support livestock grazing. Accordingly, water sources and fencing accompanied the restoration (Guest 1968). In 1954 the lands were transferred to the US Forest Service where they remain today, accounting for about 2% of the land in USFS jurisdiction. Though many regulations govern the use of the national grasslands, it is clear that the general intent is for sustainable yield; that is, these areas are to be managed under "sound and progressive principles of land conservation and multiple use, and to promote development of grassland agriculture and sustained-yield management of the forage, fish and wildlife, timber, water, and recreation resources" (Olson 1997).

After water stress, grazing is the most critical regulatory factor in the shortgrass prairie community. Prior to the arrival of European Americans, the prairie existed under grazing pressure from bison, elk, and pronghorn as well as rodents, birds, and invertebrates. One notable species, the prairie dog, was responsible for much of the structure and trophic function of the prairie ecosystem it inhabited. Relatively few plant species are tolerant of such grazing stress, and consequently endemic plant species are few in number. Much of the shortgrass prairie region was once dominated by two species, blue grama and buffalo grass, interspersed with a few other drought-tolerant grasses and forbs. The remnant prairies are not topographically homogenous, however, and various other species can be dominant in particular patches. With drought and grazing as predominant stressors we may begin to understand the regime of the shortgrass prairie. Fire is another important regulator here, though the "natural" fire frequency is unclear. There is less biomass in the shortgrass and mixed prairies than in the wetter tallgrass to the east, so it is likely that fires were historically less intense and perhaps less frequent. Natural and anthropogenic fire has played an historic role in the exclusion of woody species from the fringes of dry grassland. Sagebrush, juniper, and ponderosa pine have been excluded by fire in the northern Great Plains, as has mesquite in the south (Weaver et al. 1996).

If any of these regimes are altered, the ecological community is likely to experience some sort of change. Therefore, the ecosystem management goal to restore and maintain pre-European settlement conditions necessitates management of the stress and disturbance regimes. In the words of Weaver and others (1996), "prairie managers must first establish what pre-European conditions were and must then determine what grazing and/or fire reclamation treatments will reinstate those conditions." The historic change that is easiest to see is in grazing. With large herbivores virtually extirpated from the plains, the grazing pressure on grassland remnants was released. It has since been replaced with livestock grazing. Fire, long a tool of humans on the grasslands, was suppressed to the extent possible during the era of dryland farming. It is now employed as a management tool to discourage undesirable plant species and to stimulate growth for grazing, though its extent and recurrence interval is far less than the historic fire regime (Samson et al. 2004). Climate change is less predictable and much less manageable. Some climate models have predicted a warmer, wetter climate for the central Great Plains in the next century, continuing a trend that has been observed in recent history (Ojima and Lackett 2002). Or it may get drier, it's hard to say. The influence of climate change on the biological community of the shortgrass prairie is anyone's guess, though it seems unlikely that the "pre-European" conditions of the ecosystem can be achieved and held constant if temperature and precipitation fluctuate.

In general, humans have experimented with about every possible means of managing these prairies. Grazing and fire have been used alone and in combination with fertilization, irrigation, fencing, herbicide application, and insecticide treatment. Though some of this is done in the name of wildlife management, the overarching imperative has been "maximizing forage production to provide a red-meat commodity" (Samson et al. 2004). This is protection with a purpose – it is Pinchot's conservation on the range. In this case, the use of protected prairie ecosystems for cattle grazing has an ecological justification. These prairies appeared as they did to the European American settlers partially because of grazing, and ranching seems an appropriate use in the absence of bison. But it would be incorrect to state that grazing is being used to preserve the prairie in its pristine state. Rather, the tail is wagging the ungulate: the prairie is being used to suit the cattle.

A few examples will make the point. The native bluebunch wheatgrass is easily stressed by the heavy grazing to which many of our national grasslands are subjected. Crested wheatgrass is a non-native plant from Asia that responds much more vigorously to herbivory, and as an added bonus it is palatable to and nutritious for cattle (Savage et al. 2004). Unfortunately, it out-competes some native species, including bluebunch wheatgrass. Now, if our purpose was the strict preservation of the native ecosystem, management schemes would likely be devised to favor native plant species. This is not the case – in fact the grazing pressure on many national grasslands favors the nonnative plant. Contrast this with the story of another nonnative plant, leafy spurge, which is not so desirable. Leafy spurge is a deep-rooting perennial that has spread quickly since its accidental introduction to North America. Through rapid growth and allelopathic chemicals, it frequently out-competes native grasses and forbs. If it were suitable for cattle forage, it might be tolerable; unfortunately, the cattle avoid it and consequently exert an even more concentrated grazing pressure on the more palatable forage. Control of leafy spurge is not easy, and a variety of techniques are in use. Mowing and fire have limited success, but herbicide application and biological control agents have shown some promise. Sheep and goats will eat leafy spurge but will not eliminate it from a rangeland. One of the most successful techniques is plowing of the leafy spurge combined with planting of Russian wildrye, pubescent wheatgrass, smooth brome, or Dahurian wildrye – all aggressive nonnative plants that are considered better than leafy spurge (Lym and Zollinger 1995). These plants are considered better because cattle will eat them. Clearly, it's all about the red meat.

And then there is the prairie dog, that much-maligned burrowing rodent of the grassland. Prairie dog colonies once covered between 100 and 250 million acres of North American grassland. They now occur on less than 2% of this land. All of the usual reasons for the decline apply: habitat loss, fragmentation, and land use changes. But the demise of the prairie dog is primarily due to a far more insidious reason – a century of federal and state eradication programs "intended to benefit the US livestock industry" (Miller et al. 1994). Millions of acres have been (and continue to be) repeatedly poisoned in the name of prairie dog eradication. This is all apparently based on the perception that prairie dogs reduce the quantity and quality of forage for livestock. Abundant evidence has been offered to the contrary, as summarized by Brian Miller and others (1994): prairie dogs have been shown to effect only a 4–7% reduction in cattle forage; there is no significant difference in the market weight of cattle grazed with or without prairie dogs; the grass of prairie dog towns is actually more nutritious and is preferred by cattle; bison and numerous other large herbivores co-existed with prairie dogs for thousands of years before human intervention. Despite this logic, warfare on the prairie dog continues today, most conspicuously in national grasslands, where this native mammal is being suppressed in favor of a nonnative mammal.

The Forest Service can hardly be blamed for their use and encouragement of nonnative species – it is their mandate. According the Code of Federal Regulations, the Forest Service "is required to maintain well-distributed habitat to maintain viable populations of all native and desirable introduced vertebrate and vascular plant species" within national grasslands (Samson et al. 2003). Words like "maintain" and "desirable" are certainly open to interpretation; in fact, it could be argued that the mandate is impossible to follow on the grounds that populations cannot be maintained indefinitely. Nonetheless, the Forest Service and associated partners have proceeded as directed. Even the prairie dog is "maintained" on 3% of national grassland area. But consider what the mandate requires of the agency. It requires the identification, delineation, and categorization of different prairie habitat types, as differentiated by geographic features, climatic variation, and land use history. It means that dominant vegetation types need to be identified on the coarse scale and that the specific requirements of threatened, endangered, or at-risk species need to be provided on the fine scale. The agency must plan the structure and function of specific communities to a detailed level: how many cattle per acre, how many fires per decade, how many prairie dog towns per unit area, the level of the water table. There are even plans concerning the maintenance of a particularly suitable successional stage for various areas.

In defense of this management scheme, it is probably the only way to approach such a mandate. It may also have a positive effect of stress reduction on some grasslands. But it is a Clementsian plan, necessitated by a Clementsian mandate. In this vision, grasslands occur in units that progress through stages to a climax. Some of those stages are desirable, while others are to be avoided. Once achieved, the desirable state should be maintained by a disturbance regime designed for resilience and stability. It is a "balance of nature" approach. There is not much room here for stochasticity, for environmental fluctuation, for individualistic ecology, or for response diversity. In the long term, when temperature and precipitation change, when species require different ranges, when fire and drought patterns change, when new species arrive and others leave, to what lengths will we go to maintain grasslands in the state we deem proper?

Prairie Restoration and Umbrella Species

The two cases considered above involve prairie restoration in different circumstances. One might be considered an effort to preserve a prairie legacy; the other has all the characteristics of a multiple-use, commodity based ecosystem conservation. Yet, as we have seen, both projects are decidedly focused on the protection of the prairie in its ideal state. Both are concerned with the achievement and maintenance of a particular community of dominant species in spite of environmental fluctuations that might select for an alternate community. Both seek to arrest succession in a particularly desirable stage. Both use disturbance as a stabilizing force and seek to minimize any disturbance that might alter community structure. Ecosystem integrity, health, stability and resilience are important to both management plans. Now, setting criticisms of holism temporarily aside, I will observe that these are among the premier grassland protection projects in the nation. The ecosystems they protect and the methods by which they are protected are models for many other grassland preservation and conservation projects. And there are many other such projects. Due presumably to the perception of ecological legacy, the historic devastation of grasslands, and the growing interest in grassland habitat, prairie restoration and preservation has been undertaken by a host of private, public and non-profit organizations, as well as individual landowners and land trusts. What conceptual models inform and regulate all of this prairie restoration?

At the risk of generalizing a wide diversity of prairie protection efforts, let me suggest that many such projects have the well-being of a particular species in mind. This is the umbrella species approach. Initially, the umbrella species was conceived according to extent of its range (Wilcox 1984). The idea is this: if the protected area is large enough and of sufficient quality to provide suitable habitat for the umbrella species, it will also meet the needs of other species in the community. In some cases, the concept has been extended to a more indirect association. For example, as we saw in Chap. 1, the Karner blue butterfly has specific habitat requirements. Even though its physical range is small, protection of habitat that is suitable for this species will necessarily protect the desired community. The conservation advantage of the umbrella species approach is obvious: it simplifies the goals and mission of the project. The complexities of successional change, disturbance and stress regimes, three dimensional structure, and system function are streamlined when the requirements of one species are paramount. The danger, of course, is also in the simplification. Strict adherence to the preservation of the umbrella species can reduce a stochastic and chaotic system to a deterministic target.

Prairie birds are popular umbrella species. The North American Grouse Partnership, for example, has a specific mission: to promote the conservation of grouse and the habitats necessary for their survival and reproduction. Their 2007 Grassland Conservation Plan for Prairie Grouse is an impressive strategy to identify 65 million acres of historic and current prairie for restoration and grouse conservation (Vodehnal and Haufler 2007). Three of the 12 species of grouse that occur in North America – the sharp-tailed grouse and the greater and lesser prairie chicken – are the focus of this conservation work, and they make good umbrella species. These birds require a large and complex habitat that was historically abundant but is now severely limited. Unlike many prairie species, grouse habitat requirements are well known, and protection of these habitat characteristics are likely to provide habitat for other prairie-dependent species, particularly birds. Grouse have the additional advantage of being familiar to bird watchers and hunters as a kind of icon or flagship species of the prairie.

For a person who loves grassland ecosystems, it's hard not to like the Grouse Partnership's vision or the methods by which they identify appropriate sites for grouse conservation. Implementation of the plan would be transformative on a continental scale. It would also require an unprecedented level of ecosystem management. Much of the conservation land is identified based on its historic (meaning pre-European settlement) conditions. These conditions were primarily the result of historic fire regimes, including fires intentionally set by Native Americans, grazing by enormous herds of bison, and historic climate patterns. If, in some pro-grassland future America, the area proposed for grouse conservation could actually be acquired, it would not automatically revert to suitable grouse habitat. The "grouse commons" would first have to become a "buffalo commons" (Popper and Popper 1987), complete with re-establishment of native plant and animal species and recurrent fire. Even if this could be accomplished, the future climate may well shift the historic range of prairie organisms. In fact, it has already. My point is not that large-scale grassland protection and restoration should not be undertaken. Rather, I'm suggesting that ecosystem restoration need not provide specific historic conditions for particular species to be considered successful or beneficial.

A good example of protected land with variable successional outcomes is the Conservation Reserve Program (CRP). A component of the 1985 Food Security

Act, CRP has resulted in the conversion of about 35 million acres of marginal croplands to grassland habitat over its first 20 years. This is not quite the 65 million acres that the Grouse Partnership envisions, nor is it all strategically located to maximize grouse habitat suitability. Nevertheless, about 25 million acres of CRP land are in states that support populations of sharp-tailed grouse and greater and lesser prairie chicken. Has CRP provided grouse habitat? It has, though the response of grouse species has not been uniformly positive over the entire area. The success of individual species is a function of CRP land area, location, proximity to other grasslands, grass stand height and diversity, and management practices. Particular grouse species in some CRP states have achieved no discernable increase in range or population, while grouse populations on other CRP land have responded with vigor (Rodgers and Hoffman 2005). Should CRP be therefore considered only moderately worthwhile? Absolutely not. Of course CRP lands provide a wide range of habitat – each piece of CRP land is a unique ecosystem. If left to succession, each will develop according to its own biotic and abiotic regimes. And whether or not grouse happen to be present, each provides habitat for some biological community. Each site also contributes to the reduction of soil erosion, to carbon sequestration, and to the quality of adjacent aquatic ecosystems. The umbrella species, the indicator species, the flagship species – all are useful concepts, but we must allow for other possibilities.

In rare cases, umbrella species are not even native to the ecosystem they represent. The pheasant, for example, was introduced to the United States as a game bird in the mid-nineteenth century. An Asian native, it has been bred and introduced around the world. Pheasant hunting is so popular in the United States that the bird has become a symbol of its habitat. It is the state bird of South Dakota. (The same state that chose this nonnative species as its icon, incidentally, destroyed the largest remaining prairie dog town in the nation in the 1980s, as described by Miller, Ceballos and Reading [1994].) The pheasant is so popular that it has a conservation organization dedicated to its well-being. Pheasants Forever is a land protection organization of over 120,000 members dedicated "to the conservation of pheasants, quail and other wildlife through habitat improvements, public awareness, education and land management policies and programs." Since its formation in 1982, Pheasants Forever has played a part in the acquisition of over 100,000 acres and the habitat improvement of over 5 million acres of North American land. Far more than a hunting organization, Pheasants Forever is deeply involved with land conservation advocacy and in re-connecting people – especially children – with the land.

Pheasants Forever is Leopoldian to the core; it advances Leopold's land ethic, including his willingness to modify land to suit particularly desirable species, which in this case he called "pheasant planting" (Leopold et al. 1999). Pheasants Forever has done some fine work, and I don't wish to disparage it in any way. But let me make two observations. First, it doesn't seem to bother anyone that the pheasant is a nonnative species. On the contrary, the poor response of pheasant to CRP has been the cause of some consternation in the conservation community, resulting in proposals for alternate CRP land management to increase pheasant habitat suitability (Rodgers 1999). This makes the call to restore native grassland ecosystems to their historic condition seem somewhat hollow. Second, the pheasant, like the grouse, is an icon of its habitat because of its game-bird status. There is a reason that the organization is not called Weasels Forever or Meadowlarks Forever. Pheasants are fun to shoot. What this means, of course, is that land protection and modification for game bird habitat is yet another example of that great driver of ecosystem management: human edification.

Land Trust Grasslands

Say what you will about the origins of its icon, but Pheasants Forever has undeniably involved a great number of private landowners in the conservation movement. The primary branch of the parent organization that does this work is called the Forever Land Trust. Like the thousands of other land trusts in the country, Forever Land Trust assists landowners who are interested in donating their land and/or placing it under a conservation easement. The easement restricts the use of the land for a specified period of time. Restrictions are determined according to the wishes of the landowner with the assistance of the land trust. In the case of grassland, for example, the landowner might donate the development rights but retain the right to farm or graze the land. In a more restrictive easement the landowner could choose to relin-quish all rights of agriculture and development, retaining only rights of access and hunting. For a growing number of Americans, land trusts are a way of protecting the land that they love for the foreseeable future – even forever, if they wish, for many easements are agreements in perpetuity. According to the Land Trust Alliance, there are currently over 1,700 land trusts in existence that collectively protect over 37 million acres of land.

The land trust is an instance where ecosystem protection meets tax law, and this deserves a slight digression. A landowner who voluntarily surrenders land use rights may claim this as a charitable donation. Consequently the Internal Revenue Service has a definite interest in land conservation. If one seeks tax benefits, not just any old piece of land is worthy of conservation. To simplify a lot of complicated language, land qualifies for an ecological conservation purpose if it is "a significant relatively natural habitat in which a fish, wildlife, or plant community, or similar ecosystem normally lives" and if the conservation purpose is "protected in perpetuity" (Treas. Reg. § 1.170A-14(d)(3)(ii)). As we might expect, the definitions of "significant" and "relatively natural" are open to some interpretation. "Significant" habitat, for example, can mean that the habitat supports rare, threatened or endangered species, that it represents high quality aquatic or terrestrial communities, or that it contributes to the ecological viability of a nearby conservation area. This is all very much in keeping with our national concept of the ideal ecosystem. The notions that there are places where certain communities *normally* live, that some communities are *higher quality* than others, and that we have the means to protect such conditions in *perpetuity* speak volumes about our perception of ecosystems as static and sessile units.

Some land trusts protect grasslands, and in two such examples we can see both the umbrella species concept and management for the ecosystem ideal in practice. The Southern Plains Land Trust, for example, protects shortgrass prairie ecosystems in Colorado. Their umbrella species is the prairie dog, that keystone of pre-European American settlement prairie structure and function. Justifiably dissatisfied with government-sponsored grassland management, the Trust has set out to acquire and manage their own. Their intentions are to restore vegetation through seeding, burning, and mowing, and to reintroduce animals like the prairie dog and bison, but then to allow the ecosystem to develop with minimal human intervention. The land trust holdings are adjacent to national grasslands; the comparative succession will be interesting. Another private conservation organization, the Whidbey Camano Land Trust of Washington, has taken the bold step of removing trees from one of its properties to restore grasslands and protect its umbrella species: the state-endangered and federally threatened golden paintbrush plant. These grasslands, like those of the Great Plains, are artifacts of Native Americans whose fires cleared the land for cultivation. Removal of the offending conifers in favor of more desirable species was deemed "essential to the health of this prairie parcel and to the long-term viability of the threatened Golden Paintbrush" (Whidbey Camano Land Trust 2008). The expectation is that the grassland and its critical habitat can be restored and maintained. This will clearly be a fight against succession.

America's grasslands, from tiny demonstrations to unique remnants to vast rangeland, are largely artificial manifestations of human perception. Where preservation of historic legacy is desired, we are fighting a never-ending battle against prevailing stress and disturbance regimes, hoping to recreate the regimes and communities of the past. In grasslands that provide a useful commodity we have altered stress and disturbance regimes to serve that purpose. In general, the goal of our grassland management is to get the ecosystem to behave as we think it should: to harbor only the proper species, native or not; to resist change; to remain stable at some desirable successional stage. When disturbance is employed as a management tool it is often intended to stimulate a specific response, not to encourage general response. Of course, none of this is new. Humans have been creating, manipulating, and harvesting North American grasslands since glacial retreat. The difference, I think, is that now we hold some ecosystem states in higher regard than others due to their integrity, health, stability, and capacity for a very specific resilience. If the maintenance of our grasslands in the idealized state is truly our goal, we face a protracted struggle against contrary environmental conditions.

References

Anderson, R. 1973. The use of fire as a management tool on the Curtis Prairie. Proceedings, 12th Tall Timbers Fire Ecology Conference, June 8–9, 1972, Lubbock, TX. Tallahassee: Tall Timbers Research Station.

Cottam, G., and Wilson, H. 1966. Community dynamics on an artificial prairie. Ecology 47:88–96.

Curtis, J., and Partch, M. 1948. Effect of fire on the competition between blue grass and certain prairie plants. American Midland Naturalist 39:437–443.

Guest, B. 1968. The Cimarron National Grassland: a study in land use adjustment. Journal of Range Management 21:167–170.

Hurt, R. 1985. The national grasslands: origin and development in the dust bowl. Agricultural History 59:246–259.

Kucharik, C., Fayram, N., and Cahill, K. 2006. A paired study of prairie carbon stocks, fluxes, and phenology: comparing the world's oldest prairie restoration with an adjacent remnant. Global Change Biology 12:122–139.

Leopold, A. 1988. American eye: what is an arboretum? The North American Review 273:10–13.

Leopold, A., Callicott, J. B., and Freyfogle, E. T. 1999. Aldo Leopold: For the Health of the Land: Previously Unpublished Essays and Other Writings. Washington: Island Press Shearwater Books.

Lym, R., and Zollinger, R. 1995. Integrated management of leafy spurge Circular W-866. Fargo: North Dakota State University Extension Service.

Miller, B., Ceballos, G., and Reading R. 1994. The prairie dog and biotic diversity. Conservation Biology 8:677–681.

Ojima, D. S., and Lackett, J. M. 2002. Preparing for a Changing Climate: The Potential Consequences of Climate Variability and Change – Central Great Plains. Report for the US Global Change Research Program. Fort Collins: Colorado State University.

Olson, E. 1997. National Grasslands Management: A Primer. Natural Resources Division, Office of the General Counsel. Washington: United States Department of Agriculture.

Popper, D., and Popper, F. 1987. The Great Plains: from dust to dust. Planning 53:12–18.

Pyne, S. J. 1997. Fire in America: A Cultural History of Wildland and Rural Fire. Seattle: University of Washington Press.

Rodgers, R. 1999. Why haven't pheasant populations in western Kansas increased with CRP? Wildlife Society Bulletin 27:654–665.

Rodgers, R., and Hoffman, R. 2005. Prairie grouse population response to conservation reserve grasslands: an overview. In The Conservation Reserve Program–Planting for the Future: Proceedings of the National Conference, June 6–9, 2004: Fort Collins Science Center Scientific Investigations Report 2005–5145, ed. Allen, A. W., and Vandever, M. W., pp. 120–128. Fort Collins: United States Geological Survey.

Samson, F., and Knopf, F. 1994. Prairie conservation in North America. BioScience 44:418–421.

Samson, F., Knopf, F., McCarthy, C., Noon, B., Ostlie, W., Rinehart, S., Larson, S., Plumb, G., Schenbeck, G., and Svingen, D. 2003. Planning for population viability on Northern Great Plains national grasslands. Wildlife Society Bulletin 31:986–999.

Samson, F., Knopf, F., and Ostlie, W. 2004. Great Plains ecosystems: past, present, and future. Wildlife Society Bulletin 32:6–15.

Savage, C., Williams, J. A., and Page, J. R. 2004. Prairie: A Natural History. Vancouver: Greystone Books.

Vodehnal, W. L., and Haufler, J. B. 2007. A Grassland Conservation Plan for Prairie Grouse. Fruita: North American Grouse Partnership.

Weaver, T., Payson, E. M., and Gustafson, D. L. 1996. Prairie ecology – the shortgrass prairie. In Prairie Conservation: Preserving North America's Most Endangered Ecosystem, ed. Samson, F., and Knopf, F., pp. 67–75. Washington: Island Press.

Whidbey Camano Land Trust. 2008. Land trust begins prairie restoration in Ebey's Reserve. Press Release, April 2008.

Wilcox, B. A. 1984. In situ conservation of genetic resources: determinants of minimum area requirements. In National Parks, Conservation, and Development: The Role of Protected Areas in Sustaining Society. Proceedings of the World Congress on National Parks, Bali, Indonesia, October 11–22, 1982, ed. McNeely, J. A., and Miller, K., pp. 18–30. Washington: Smithsonian Institution Press.

Chapter 9
Freshwater Ecosystems

If the history of widespread destruction and subsequent urgency of preservation in America's grasslands is rivaled by any other ecological environment, it is our freshwater ecosystems. For three-fourths of its history, the United States has been devoted in policy and practice to wholesale abuse of rivers, lakes, and wetlands. Structural modifications to aquatic systems during this phase of American history were commonly called "improvements," putting a positive spin on drainage, impoundment, channel straightening, floodwalls, dams, and dredging. In recent decades the policy has come about face, so that certain measures of protection have been extended to freshwater systems. Compliant practice has generally involved an un-doing of the previous improvements. Thus we are un-damming, re-meandering, re-flooding, re-vegetating, and un-diverting across the nation. These new improvements have not yet come close to the scale or pace of the original modifications, but after a solid 50 years of freshwater ecosystem restoration we have enough examples to know what can and what can't be easily restored.

It is comparatively easy to restore and maintain the *structure* of aquatic ecosystems, for it is an engineering question of getting the water to go where you want it to go. Restoration of the ecosystem's composition and function is another matter. Living things do not always do what we want them to do. And yet this is often the implicit goal of ecological restoration, conservation and preservation, particularly when the ecosystem has a legacy or a commodity that is important to humans. As a result, America's efforts to protect and restore its rivers, lakes and wetlands are often shaded by a desire to get the ecosystem to behave as it once behaved, to function as we think it should, and to remain unchanged.

Wetlands

Freshwater wetlands have undergone an impressive image makeover though the course of American history. At the time of European settlement, there were an estimated 220 million wetland acres in the conterminous United States (Dahl 1990). At the time they were almost universally reviled as impediments to travel, agriculture, human health and progress. American policy throughout the nineteenth and well

D. J. Spieles, *Protected Land*, Springer Series on Environmental Management,
DOI 10.1007/978-1-4419-6813-5_9, © Springer Science+Business Media, LLC 2010

into the twentieth century was intended to encourage the drainage of every swamp, marsh, pothole, and bog in the country – and it was nearly successful. For instance, consider the Swamp Land Acts of the 1850s, which deeded wetlands to states for the purpose of drainage and cultivation. Nearly 65 million acres of federally owned wetlands were disposed of in this manner, with most ultimately coming under private ownership. By 1980, about half of the wetlands in the conterminous Unites States had been drained or filled (Lewis 1995). But not everyone accepted this massive drainage effort as an improvement. In the first half of the twentieth century there was growing concern for, and even legislation to protect, migratory waterfowl. Organizations like the Izaak Walton League (founded 1922), Ducks Unlimited (1937), and various state-level "Save the Wetlands" programs were formed in part to halt the destruction of wetlands, even as the government continued to subsidize drainage. By the 1970s, the national attitude began to shift, and with the 1977 amendments to the Clean Water Act wetlands came under federal protection.

From this schizophrenic history was born one of the strangest and most convoluted mandates of ecosystem conservation in American history. Here is the abridged version: The Clean Water Act and its interpretations do not prohibit the destruction of wetlands, they just require that the individual or organization planning the destruction obtain a permit, usually from the United States Army Corps of Engineers. The permit process requires the permittee to demonstrate that the proposed activity cannot avoid wetland encroachment and that the damage has been minimized to the greatest degree possible. With these formalities met, the overwhelming majority of permits are approved, provided that the permittee agrees to compensate for the loss of the wetland by constructing, restoring, or enhancing a wetland in another location. This is called wetland mitigation, and it is the heart of the national policy called "no-net-loss." The idea is simple on paper – for each acre of wetland unavoidably lost to development, one or more acres of wetland will be created in a more convenient location. In theory, it is a win for all parties: development continues unimpeded, ecosystem services remain intact, and wildlife habitat is maintained or even increased. In reality things are a bit stickier. The legal and political complications are numerous and entertaining but beyond the scope of this work, so regarding questions of agency jurisdiction, which wetlands apply, how wet they have to be, how connected or large they have to be, which wetlands and conditions are exempt, and the various ways in which the mitigation obligation may be met the reader is referred elsewhere (Lewis 2001).

Even without these scintillating topics, the no-net-loss policy provides rich fodder for our discussion. Let me reduce it to just two questions: First, by what criteria do we evaluate the suitability or "success" of mitigation wetlands? Second, are these criteria meaningful in the absence of regulatory structure? These are real and important questions of national wetland conservation policy. They are also representative of broader issues of ecosystem conceptualization and management. In essence, they are at the heart of our search for a yardstick with which we can evaluate ecosystem development.

Wetland Creation and Restoration

Mitigation obligations are most commonly met by wetland creation, which is the construction of a wetland in a place where no wetland formerly existed, or by wetland restoration – the re-establishment of a particular wetland where it once occurred. In each case, certain ecosystem performance standards are established to ensure that the constructed wetland is a reasonable replacement for the wetland that was lost. Typically, characteristics of the wetland plant community are one – if not the only – performance standard. The plant community is seen as a surrogate for ecosystem structure and function; the thinking is that if the proper plants are present, then the conditions of the mitigation wetland must be suitable. Now, the next logical question might be this: which are the proper plants? There is no single answer, but in general vegetation performance standards require a plant community that is (1) predominantly native; (2) composed primarily of wetland plants; and (3) of reasonably high quality. The native concept we have encountered in other ecosystems. The desire for wetland plants is simply based on the observation that some plants are better adapted than others for life in standing water. For instance, a mitigation wetland that supports only upland plants is no good at all – only the establishment of obligate or facultative aquatic species would indicate that a wetland exists.

What about the third criterion – establishment of a high quality plant community? This is a measure of biological integrity, meaning that certain species are deemed more valuable or desirable than others. What, then, are the characteristics of a desirable wetland plant community? One way this is determined is by computation of the *floristic quality* of the community – a scoring system that has been developed for many areas of the United States. A floristic quality index assigns a quality value to various species based on the specificity and tolerance they exhibit in reference habitats. For example, nonnative plant species and aggressive, opportunistic natives are awarded zero points, native species that are relatively tolerant to stress or disturbance might be awarded a medium score, and native species that are intolerant to all but the most unstressed and narrow conditions would get the highest score (Cronk and Fennessy 2001). Thus, a created or restored wetland that is able to establish a plant community with high floristic quality in a short time (generally 3–10 years) has a good chance of meeting regulatory approval and fulfilling the mitigation obligation.

Of course, highly specialized and intolerant species are characteristic of late-successional ecosystems, while tolerant and opportunistic species are common in early succession. The mitigation game, then, is to establish the physical structure of the replacement wetland, introduce thousands of high quality plant species, and keep them alive until regulatory approval. This also means that early successional species must be discouraged (even though the ecosystem is in fact early successional), lest they out-compete the high-quality species, reduce the integrity score, and delay approval. In effect, the regulatory structure means that the mitigation wetland planners and managers must envision a late successional target ecosystem and then use every weapon in their arsenal to achieve that end as rapidly as possible.

There are far too many examples of this assembly line approach to wetland construction. Here we will consider only one (New and Associates 2003). The Lake Station mitigation wetland in northern Indiana was constructed in 1998. The 200 acre site was formerly agricultural land, and the goal was to restore the wet prairie that one existed. Drainage tile was removed and impoundments were constructed to retain surface water. The site was mowed and sprayed prior to seeding and introduction of 30,000 wetland plant seedlings. Undesirable weedy species were persistent, so the wetland was burned in the dry winter of 1999. Mowing and spraying continued in 2000, along with specific control of invasive *Phragmites*, purple loosestrife and encroaching woody species. Loosestrife beetles were introduced as a biological pest control agent. More plantings and weed control continued in 2001, particularly to beat back woody species. By 2002, *Phragmites*, purple loosestrife, thistle, and woody invasive were still problematic, and the site's hydrologic regime was not sufficiently supporting the desired wetland plant community. So, the site was burned again and excavated to improve the hydrology and to encourage the survival of planted species. By this time, however, management was moot in a regulatory sense, for enough appropriate vegetation had survived for regulatory approval in 2001.

My purpose is not to suggest that the Lake Station wetland is inappropriate habitat, nor do I relate this information to belittle the efforts of restoration ecologists. Instead, I share this example to illustrate the inherent flaws in our process of mitigation. The rapid establishment and maintenance of a late-successional plant community in an early successional environment is entirely contrary to way ecosystems assemble and change. The selection of desirable species and engineering of abiotic conditions to perpetuate those species makes an artificial ecosystem that is sustainable only as long as humans can keep burning, mowing, spraying, and manipulating water levels. It has been called the "designer" approach to ecosystem restoration, meaning that the result is an ecosystem conceived and constructed by humans (Mitsch and Gosselink 2007). We would have to place this at the extreme of the holistic idealism continuum, right alongside sanctuary preservation.

There are other approaches to wetland creation and restoration that are not quite so prescribed. *Self-organization* is the idea that systems assemble themselves and achieve a stable state based on inputs and constraints (Odum 1989). The manager, then, needs only to establish and maintain the proper conditions and the community will assemble itself. This management scheme is called self-design, and it is based on the assumption that "the system will optimize its design by selecting for the assemblage of plants, microbes, and animals that is best adapted to the existing conditions" (Mitsch and Gosselink 2007). This expectation of "self-optimization" might smack of the superorganism to the radical individualist, who might argue that there is – or should be – no design in these systems at all, just a coincidence of species individualistically exploiting resources and one another in a fluctuating framework of stress and disturbance. But one would be foolish to subscribe to the "self-design" or "no-design" approaches to mitigation if regulatory approval – and a great deal of money – is on the line in just a few years. The "designer" approach ensures swift assembly of the ideal community and the greatest likelihood of approval. It is politics and economics, not ecology, which drives ecosystem management in this case.

Actually, the "designer" approach is frequently taken even in wetland creation and restoration projects that are not for compensatory mitigation. The reason is that, after several decades of wetland construction, many of the "no-design" wetlands have yielded unsatisfactory results. For example, thousands of small wetlands have been created or restored in the prairie pothole region of northern Iowa, western Minnesota, and North and South Dakota, primarily for the improvement of regional water quality and waterfowl habitat (Galatowitsch and van der Valk 1994; Mulhouse and Galatowitsch 2003). Historically rich in wetlands, this area has been extensively drained for agriculture. The wetland restoration methods have typically been simple, consisting of the removal of drainage ditches and tile and the expectation of seasonal flooding and natural revegetation. In a study that assessed 64 of these restored potholes 3 years after construction, it was found that wetland plant species quickly revegetated the sites, either by dispersal from afar or by seedbank remnants (Galatowitsch and van der Valk 1995). In only 3 years, aquatic perennials and submerged and floating plants had colonized areas of consistent standing water, though fringe areas had not yet returned to wet prairie or sedge meadow habitats. Twelve years after construction, species richness had increased but the wetlands were in general dominated by a few invasive perennials (Mulhouse and Galatowitsch 2003). The characteristic vegetation zonation – the ecosystem structure – of a typical pothole had returned in most cases, though the wet prairie and sedge meadow still had not developed. This, it seems, is the result of the "no design" approach: an ecosystem that approximates the intended structure but that lacks the species composition of the target community. As a result, there is a growing reluctance to rely on "natural" restoration: "Given the dominance of invasive perennials and the absence of many native wetlands species, it appears that without significant seeding, planting and aftercare wetland restorations in fragmented landscapes have a low probability of resembling those that existed historically" (Mitsch and Gosselink 2007).

The trepidation for individualistic, "no-design" restoration is drawn from the fear that monotypic stands of aggressive perennials will preclude the colonization of native plants that once inhabited the ecosystem. Indeed, there is a great deal of evidence that this can and does happen, but it is less clear why exactly it is a problem. There is, somewhat ironically, concern that the invasive dominants will not be easily displaced, thereby arresting succession in an undesirable state. There may also be a trend toward homogenization, if all restored wetlands are overcome by these same few dominant species. At present, the restored potholes described above are not utterly devoid of plant diversity, but the diversity is less than, and different than, comparable natural wetlands. Function is even more difficult to evaluate. The restored wetlands retain floodwaters, improve water quality, and provide wildlife habitat. Do they perform these functions in a way that is consistently equivalent to the historic wetlands of the prairie pothole region? Maybe not. But they are as they are because of the environmental matrix in which they were constructed. They are a product of their drainage history, soil characteristics, flood regime, and proximity to potential colonizers. We may choose to view their development as succession gone awry, as some failure of resilience. But these ecosystem are not failures of self assembly, they are collections of individual species responding to the stressors and opportunities of their environment. I'm not convinced that forcing a designer ecosystem into a landscape that no longer supports it is a better option.

Lakes

The prairie pothole restoration scenario demonstrates what some might call a *regime shift*. Natural potholes, the theory goes, existed prior to human encroachment in a particular structural and functional state. Radical changes to disturbance and stress regimes and to the species pool have rendered the former state untenable, and an alternate state has emerged. The restored wetlands thus exist in a new "domain of attraction" and may never revert to the former state. The regime shift is intimately linked to the idea of resilience. For example, a natural pothole may be relatively unchanged by some minor disturbances, like duck hunting or a brief drawdown. The ecosystem is seen as resilient to these small alterations. But increase the pressure – as in extensive drainage or vegetation removal – and you might reach the threshold of a new state which will not achieve the former species composition in any amount of recovery time. Regime shifts occur naturally and can certainly be caused by humans. It is the rate at which humans are altering ecological regimes and the consequences for ecosystem services that are seen as cause for alarm (Folke et al. 2004).

Regime shifts have been documented in a wide variety of ecosystems, but the cause, threshold and effect might be the most demonstrable in lakes. Lakes appear to be rather fixed features of the landscape, but like all ecosystems they are temporary over geologic time. And over time, they change. Though obviously unlike terrestrial or even wetland ecosystems, lakes support a particular trophic structure and biogeochemical function, both of which undergo succession. The classical view of lake succession is based on trophic status (Horne and Goldman 1994). It is based on the assumption that most lakes, upon formation (by glacial activity or geologic events, for example), are nutrient poor. These oligotrophic environments, characterized by low productivity and relatively high oxygen availability, progress through weathering, runoff, erosion, biological activity, and sedimentation to a mesotrophic state in which nutrients are more plentiful within the system. As productivity increases more nutrient fixation occurrs within the system, driving it toward a nutrient rich eutrophic state. Gradual sedimentation and detritus accumulation over time would lead to a shallow, decomposer-dominated lake, which would eventually fill in to the point where it might be better described as a swamp or bog. Ultimately, the lake would cease to exist and become a terrestrial ecosystem.

While this scenario may occur, it is clearly not the only successional progression for lake ecosystems. The rates of sedimentation and organic accumulation are highly dependent on the mechanisms of lake formation, the geomorphic setting, and the fertility of its drainage basin. As these factors are different in every situation, there are no "typical" stages of lake succession or common endpoint. Some lakes are shallow formations that receive rapid deposition and might fill in after a few thousand years, while others will likely not fill in before they are altered or destroyed by the next cataclysm. Even the generalization that lakes proceed from nutrient poor to nutrient rich conditions is not always accurate, for many lakes have historically experienced periods of high productivity prior to periods of nutrient scarcity. If we are able to make any general statements about lake ecosystems, it would be that (1) lakes change over time in both a biotic and an abiotic sense; (2) successional processes in lakes are not goal oriented; and (3) lake trophic state and successional progression is highly linked the status of the lake's watershed (Wetzel 2001).

In developed nations like the United States, another fair generalization would be that human activity tends to drive lakes toward nutrient enrichment. Cultural eutrophication is most commonly the result of wastewater discharge, runoff of terrestrial fertilizers, and atmospheric deposition. Eutrophication is a natural lake process, but this anthropogenic eutrophication is accelerated and intensified. The effect of sudden eutrophication on a lake can constitute a regime shift. An all-too-common regime shift scenario in temperate lakes goes something like this: a lake is relatively nutrient poor, largely due to low inputs from its forested watershed. As an oligotrophic system it has clear water, low phytoplankton biomass, a low amount of organic detritus accumulated in its bottom waters, and relatively plentiful dissolved oxygen. Enter the human culture, which clears and drains the watershed for agriculture, applies fertilizer to the land, and discharges high-nutrient waste into the lake and the rivers that feed it. Now the lake experiences algal blooms and reduced clarity. The nutrient inputs, particularly of phosphorus, are recycled repeatedly within the lake and accumulate with continued loading. The bottom waters accrue oxygen-demanding detritus, and the ensuing hypoxia leads to fish kills. Vascular plants that prevented erosion and stabilized sediments in shallow zones are out-competed by algae, and in their absence sediments are easily resuspended and the lake becomes more turbid. Cyanobacteria invade, releasing toxins and fixing atmospheric nitrogen, which further hastens eutrophication. With some variation, this clear-to-turbid regime shift is a common consequence of modern human culture (Folke et al. 2004).

From a human perspective, it would be hard to argue that the above scenario is anything but a great loss of ecosystem services. The lake once supported game fish, and now those species are rare or gone. Clean drinking water, desirable conditions for recreation, and benefits to adjacent terrestrial or wetland ecosystems may also be reduced. Equally troubling is the observation that the turbid state, once achieved, is not easily returned to the clear state by nutrient diversion or upland erosion prevention. The scenario is often perceived as an either-or, with two alternate stable states, one desirable and one not. Coupled with the fact that many of our lakes have undergone cultural eutrophication, this has given rise to a wide variety of methods intended to return a lake to its desirable state (Spieles 2005). Aside from nutrient diversion, lake managers have attempted with varying success to immobilize the phosphorus in lake sediments by massive addition of aluminum, iron, or calcium salts; they have also oxygenated the lake, dredged the bottom to remove phosphorus-rich sediments, and removed water from the lake bottom to reduce internal phosphorus loading. None of these physical or chemical methods will restore the desired biotic community, so it is accompanied by biomanipulation: fish elimination to encourage vascular plant growth and reduce turbidity, stocking of piscivores to replenish gamefish populations and to reduce planktivorous fish, and habitat modification to encourage the reproduction of desired species. The result may be something that resembles the "clear state," though if the conditions of eutrophication remain the management can be required indefinitely.

Alternate States and Desirability

It is not quite so simple to view lakes – or any other ecosystems – as having only two alternate states. Even proponents of alternate state theory acknowledge "multiple basins of attraction" (Folke et al. 2004). This is easily demonstrable in the clear and turbid regime shift scenario. Lake Erie is a textbook case of cultural eutrophication and regime shift. Erie is the tenth largest lake in the world by surface area but the shallowest of the Great Lakes. Its morphology and setting make it susceptible to change. Historically a clear lake with abundant populations of piscivorous gamefish, the water quality underwent major changes beginning with European American settlement in the mid-nineteenth century. All of the agents of cultural eutrophication occurred here in spades, and within a century the lake was an algal bowl (Vallentyne 1974). A regime shift had occurred, and for a time the lake was essentially a turbid, hypereutrophic, and oxygen-starved ecosystem.

The recovery was initiated by the Great Lakes Water Quality Agreement signed by the United States and Canada in 1972. This and other legislation reduced Erie's nutrient load by addressing point sources, notably sewage and industrial effluent. Gradually, the Lake has undergone *oligotrophication*, and its clarity has improved. The clarification was assisted in the late 1980s by the accidental introduction of the zebra mussel, a prolific filter feeder. The algal turbidity decreased over the next decade with a rapidity that cannot be explained by nutrient diversion alone – in fact clarity has improved even in spite of a slight increase in phosphorus loadings (DePinto and Narayanan 1997; Ludsin et al. 2001). Zebra mussels have apparently caused a regime shift of their own. Now we have a situation in which Lake Erie is remarkably clear, and game fish and vascular plant populations have rebounded – and yet there is still a phosphorus and detritus-laden sediment that causes seasonal oxygen depletion. Both conditions, the clarity and the sediment phosphorus load, are thanks to the industrious zebra mussel. And so which regime – which basin of attraction – is this? It has components of both the clear state and the turbid state, but it is not exactly like either. It is a third alternate state. The third of how many?

Now consider another lake – also the tenth largest in the world, but in this case according to depth. Lake Tahoe is an alpine lake of the Sierra Nevada range, and *its* morphology and setting have long protected it from eutrophication. Tahoe is not immune to the pressures of eutrophication, for despite sewage diversion its clarity has decreased in the past half-century due to erosion from human development and atmospheric deposition of nitrogen (Goldman 2000). Still, it does not appear that the lake has passed the threshold to a eutrophic regime.

But there are other possible alternate states. Tahoe's regime shift has been food web-based. In 1888, lake trout were introduced as a game fish. In the 1950s, over 11 million Kokanee salmon fry were introduced for the same reason. To support these recreationally valuable species, the opossum shrimp, a native of eastern North America, was introduced as a food source in the mid-1960s. All three species are firmly entrenched in the Tahoe food web today (Richards et al. 1975). Their presence has not been innocuous. Opossum shrimp are planktivores, with a selectively voracious appetite for cladocerans like *Daphnia*. Since shrimp introduction, cladoceran numbers have declined dramatically, as has their predation pressure on phytoplankton. Meanwhile, the introduced trout and salmon, augmented by the introduced shrimp, grow large enough to prey upon other planktivorous fish – thus native populations of tui chub, mountain whitefish, Lahontan speckled dace, and Tahoe sucker have declined (Vander Zanden et al. 2003). These native species predominantly were benthic feeders; they have been replaced by pelagic feeders. Furthermore, the introduced trout have replaced the native Lahontan cutthroat trout as the top predator in the lake. The result of all this is a profoundly different food web – not only with different regulating organisms exerting a different pressure on lower trophic orders, but also occurring in different zones of the lake. This constitutes a regime shift.

Actually, it constitutes several regime shifts. No period in the last 150 years of Lake Tahoe history has had the same food web conformation for more than a decade or 2 (Vander Zanden et al. 2003). Trophic change in Lake Erie shows the same thing: there are many alternate ecosystem states, perhaps as many as there are permutations of biotic and abiotic variables. It is only when the change is large or disagreeable enough to attract human attention that we call it a regime shift.

Whether or not a shift is agreeable is open to some debate. To those interested in the restoration of Lake Tahoe's historic food web, this latest regime is a barrier. To purveyors and participants of the annual Lake Tahoe Kokanee Salmon Festival – including mascot Sammy Salmon – this new regime is a boon. The eutrophication of Lake Erie led to some undesirable circumstances, as the turbid state is not particularly conducive to recreation. The new "clear state" brought on by the zebra mussel is far better for swimmers and anglers, though the zebra mussel is the scourge of many other lake interests. So desirability is a matter of perspective. Consider this: the United States has worked hard to reduce lake eutrophication over the last 40 years to return our lakes to a more desirable state, but during the same time lakes in other parts of the world have been intentionally driven to a eutrophic state for aquaculture (Qin et al. 2007). It's all in the eye of the beholder. Without a doubt, there are ecosystem states that can be desirable or undesirable for particular cultures at particular times. But the idea that the native, ideal state is inherently better than any alternate is not scientifically supportable, and as a management goal it is unachievable.

Rivers

Try to apply the classic model of ecological succession to a flowing body of water and you'll quickly realize that it's a bad fit. We can't exactly say how a young river typically looks or acts; it depends entirely on its geomorphic setting and on the events that led to its formation. Likewise, we generally don't speak of riverine communities as units like we do for forests (e.g. the spruce-fir community) or other stationary ecosystems. Nor do we label some stretches of river as climax communities or old growth. This is not to say that riverine communities do not undergo succession, encounter stressors, or respond to disturbance – just that the terrestrial and lacustrine models are inadequate to describe these changes. River succession is a story of flux. The state of any given reach at any given time is a function of the material and energy flow entering it from upstream and laterally from its terrestrial corridor. Far from stable, the river ecosystem is subject to short term changes like alteration of flow rate, sedimentation, or organic inputs as well as long term changes such as channel modification and watershed dynamics. Even those characteristics that might be considered descriptive of the river community, like fish spawning sites, are constantly *moving*. The cliché is true: you never can step in the same river twice.

Successional differences notwithstanding, there are some similarities among rivers and other ecosystems regarding human management. Rivers are quite useful for the transport of goods, people, and water itself, but they have the annoying habits of flooding and meandering. For sedentary cultures this won't do at all, and so humans have a long history of amending river behavior. Straighter, more uniform rivers remove water from an area more rapidly, and since so much of our landscape is paved, roofed, and artificially drained we have increased the need for high flow capacity. Floodwaters can damage public and private holdings; to minimize this we have dredged channels and constructed floodwalls, dams and impoundments. Of course, this increases the threat of flooding downstream, so the next town must do the same. River meandering is incompatible with human infrastructure, so we have reinforced banks with concrete, rock and steel.

All of this is quite reminiscent of human modification of wetlands and lakes. And as with those systems, humans have recently recognized the errors of their ways and attempted to restore some rivers by removing flood control structures and reconstructing river habitat. Often the physical engineering (or re-engineering) of a river is accompanied by manipulation of the biota. Sometimes biomanipulation is attempted in the absence of physical modification. In any case, we have extensively attempted to reduce or eliminate nuisance species and replace them with species that we find to be more palatable. This has been accompanied by a rather impressive stocking operation, whereby millions of fish are raised in federal and state hatcheries each year to be released into streams and lakes of the United States. The hatcheries have a role in the propagation of imperiled native fish populations, but they are predominantly intended to support recreation and harvest; thus the introduction of top predators, nonnative species, species that will not survive overwinter, or even hybrids that do not occur naturally have long been standard practice. Concern for the ecological damage that might result from stocking has translated into reform only recently. Still, hatcheries persist in response to federal and state mandates to conserve native fish species and to enhance recreational fishing (Nickum et al. 2005). Both mandates, I will observe, are drawn from the school of strong holism.

The lessons we can learn from river modification and restoration concern ecosystem boundaries and the landscape matrix. Aquatic ecosystems would appear to have obvious boundaries, for the water contains the biota. Where the water ends, the ecosystem should end. But even this simple concept of an ecosystem's spatial constraint is problematic. In wetlands it is not always clear where the water ends. Lakes and rivers have more dramatic edges, but even these blend into the terrestrial world around them in ways that are sometimes gradual and seasonally variable. The actual ecosystem boundary depends upon the particular sub-system of interest. A lake's bass population is pretty clearly bounded by the water, but the lake's nitrogen cycle must include the atmosphere, influent bodies of water, terrestrial biota, and processes within the watershed. Conceptualization of the lake from the perspective of migratory bird populations would reveal yet another set of boundaries. The point is that ecosystem boundaries are arbitrary. We define the boundaries of the ecosystem by basing them on the most appealing or most important species and processes. It usually follows that our boundary definition dictates the expected structure and function of the ecosystem and the management scheme to achieve that end.

One set of river boundaries is defined by organisms and processes that occur within the flowing water. The classic river continuum concept uses this boundary definition to identify the trophic condition in rivers based on their order (Vannote et al. 1980). Low order streams, for example, are analogous to oligotrophic lakes. With labile nutrients relatively scarce, the biotic communities of these headwater streams should be structured around the processing of coarse particulate organic matter. The trophic structure of a higher order river is much more dependent on autochthonous production and on the flow of processed nutrients from upstream. The river continuum model is a useful description of longitudinal change in stream ecosystems, but it is constrained by boundaries at the water's edge. A broader view of river function might consider environments beyond the channel. In the vertical direction, river water interacts with the substrate of the stream bed; in the lateral direction the river exchanges material with its wetland or upland corridor. Thus the river is more than the biota and associated processes within the flowing water – it is a collection of subsystems for which the "boundaries are fluid and not always distinct" (Fisher et al. 1998).

Broadening and blurring the boundaries of the river allows for an understanding of ecosystem structure and function that reaches beyond nutrient availability and trophic status. It allows for an expanded notion of stress; for example, the biotic community in a particular reach might be constrained by the quality of its substrate. It is also disturbance-inclusive, for boundaries that envelop the riparian corridor necessarily involve the frequency, extent and implications of flooding. In this expanded view of a river, what we think of as ecosystem function is really an aggregate of subsystem processes that are linked but spatially and temporally separated. A change in stress or disturbance may well affect different subsystems in different ways, and each subsystem – even different patches of a subsystem – will respond uniquely. Continuing this train of thought leads one to an individualistic view of the ecosystem. If ecosystem functions are the product of linked but independent subsystem processes, then holistic qualities like resilience and health are conceptual boundaries that we draw around particular sets of individual environmental response. How, then, do we reconcile individualism and holism in river management?

River Restoration and Response

There is no shortage of river management examples from which to choose – by one estimate there are over 37,000 river restoration projects in the United States, and new work is supported by over $1 billion annually (Bernhardt et al. 2007). Unfortunately, few of these projects measure specific objectives in scientifically meaningful ways. The Juday Creek restoration in northern Indiana is a rare example of experimental design in river management (Moerke et al. 2004). The Juday is a third-order, spring-fed stream that historically has supported trout. Much of the stream was channelized in the 1950s to conform with watershed land-use changes to agriculture and residential development. The restoration involved the re-meandering, bank re-vegetation, riffle and pool construction, and erosion control of two reaches of the Juday, while a third reach upstream was left unrestored as a control. No animals were introduced; in fact, the degree to which the fish and macroinvertebrate communities could recolonize was the basis for assessment of the structural restoration. Five years after construction, success was a mixed bag. Macroinvertebrate density and fish biomass in the restored reaches had surpassed that of the unrestored, though neither fish nor macroinvertebrate diversity in the restored sections exceeded that of the control. The habitats of the restored reaches had definitely been changed – they were heterogeneous and amended with coarse woody debris, gravel, and boulders that quickly became algae-coated. Yet 5 years was not long enough to achieve significantly greater diversity in the improved habitat. Furthermore, the two restored reaches differed from one another in the recovery of indicator organisms, and the indicator organisms themselves recovered at different rates.

And so we may wonder if the restored reaches of Juday Creek have shown resilience. But doesn't the answer depend entirely on one's definition of resilience – on the physical and conceptual boundaries placed on the ecosystem? Doesn't it depend on which characteristics we expect the restored reaches to exhibit? It does indeed, and the authors of the study specified such expectations, some of which were achieved. Those restoration goals that weren't achieved can hardly be blamed on a lack of resilience; rather, they are attributable to environmental conditions beyond the channel. Even substantial modifications to an ecosystem cannot be expected to produce the ideal state in the context of an altered landscape.

Perhaps the Juday Creek example is too young or within a watershed that is too degraded to show resilience and persistence of the desirable state. Another stream alteration, in the Hubbard Brook experimental forest of New Hampshire, is an older example of stream recovery (Likens and Bormann 1995; Likens 2004). The Hubbard Brook project has yielded a wealth of data; here I consider only a few points. This forest was logged from 1910 to 1917 and then left to successional recovery. Selected watersheds were harvested again in the 1960s, 1970s, and 1980s for the purpose of long term ecological research. I will focus on stream processes in a watershed (W2) that was cut and treated with herbicide from 1965 to 1968 in comparison with a nearby watershed that has been in recovery since 1918 (W6). Given 35 years of recovery in a relatively low-stress landscape, has the stream in

W2 achieved equivalence with that of W6? In some respects it has. After a large flush of stream nitrogen immediately post-harvest, the stream of the younger watershed is statistically similar to the older watershed in terms of nitrogen content and runoff volume per unit area. However, the accumulation of large woody debris and associated phosphorus retention are still much greater in the stream of the older watershed (Warren et al. 2007). Again, it seems that an evaluation of resilience requires some careful definition. But here's the kicker: the stream of the 90-year-old reference watershed is not a stable baseline – rather, it is a moving target. The in-stream organic matter accumulation and processing are heterogeneous in space and time, species dependent, and subject to long term fluctuations. Far from demonstrating ecosystem-level recovery to a domain of attraction, the Hubbard Brook experiment has revealed "the non-equilibrium nature of these systems as they respond to regional changes in inputs" (Findlay et al. 1997).

The abuse of American freshwater ecosystems has not ceased, but it has abated somewhat in this most recent chapter of the nation's history. The abatement has been accompanied by an equally vigorous effort to restore aquatic ecosystems to a more desirable state. As we have seen, the definition of the appropriate restored state ranges widely, based on historic conditions or regulatory mandates or reference ecosystems or ecosystem service or aesthetics. In many cases, the restoration effort seeks to achieve and maintain a carefully defined ecosystem state in spite of environmental conditions and with a surprising disregard for natural colonization and successional progression. The result is holistic management thrust upon an individualistic world.

References

Bernhardt, E., Sudduth, E., Palmer, M., Allan, J., Meyer, J., Alexander, G., Follastad-Shah, J., Hassett, B., Jenkinson, R., and Lave, R. 2007. Restoring rivers one reach at a time: results from a survey of US river restoration practitioners. Restoration Ecology 15:482–493.

Cronk, J. K., and Fennessy, M. S. 2001. Wetland Plants: Biology and Ecology. Boca Raton: Lewis Publishers.

Dahl, T. 1990. Wetlands Losses in the United States 1780's to 1980's. Washington: United States Department of the Interior, Fish and Wildlife Service.

DePinto, J. V., and Narayanan, R. 1997. What other ecosystem changes have zebra mussels caused in Lake Erie: potential bioavailability of PCBs. Great Lakes Research Review 3:1–8.

Findlay, S., Likens, G. E., Hedin, L., Fisher, S. G., and McDowell, W. H. 1997. Organic matter dynamics in Bear Brook, Hubbard Brook Experimental Forest, New Hampshire, USA. In Stream Organic Matter Budgets: An Introduction, ed. Webster, J., and Meyer, J., pp. 43–46. Journal of the North American Benthological Society 16:3–161.

Fisher, S., Grimm, N., Martí, E., Holmes, R., and Jones, J. 1998. Material spiraling in stream corridors: a telescoping ecosystem model. Ecosystems 1:19–34.

Folke, C., Carpenter, S., Walker, B., Scheffer, M., Elmqvist, T., Gunderson, L., and Holling, C. 2004. Regime shifts, resilience, and biodiversity in ecosystem management. Annual Review of Ecology, Evolution, and Systematics 35:557–581.

Galatowitsch, S. M., and van der Valk, A. 1994. Restoring prairie wetlands: an ecological approach. Ames: Iowa State University Press.

Galatowitsch, S. M., and van der Valk, A. 1995. Natural revegetation during restoration of wetlands in the southern prairie pothole region of North America. In Restoration of Temperate Wetlands, ed. Wheeler, B. D., Shaw, S. C., Fojt, W.J., and Robertson, R. A., pp. 129–142. Chichester: Wiley.

Goldman, C. 2000. Four decades of change in two subalpine lakes. Proceedings of the International Association of Theoretical and Applied Limnology 27:7–26.

Horne, A. J., and Goldman, C. R. 1994. Limnology. New York: McGraw-Hill.

Lewis, W. M. 1995. Wetlands Characteristics and Boundaries. Washington: National Academy Press.

Lewis, W. M. 2001. Wetlands Explained: Wetland Science, Policy, and Politics in America. New York: Oxford University Press.

Likens, G. E. 2004. Some perspectives on long-term biogeochemical research from the Hubbard Brook ecosystem study. Ecology 85:2355–2362.

Likens, G. E., and Bormann, F. H. 1995. Biogeochemistry of a forested ecosystem. New York: Springer.

Ludsin, S., Kershner, M., Blocksom, K., Knight, R., and Stein, R. 2001. Life after death in Lake Erie: nutrient controls drive fish species richness, rehabilitation. Ecological Applications 11:731–746.

Mitsch, W. J., and Gosselink, J. 2007. Wetlands. Hoboken: Wiley.

Moerke, A., Gerard, K., Latimore, J., Hellenthal, R., and Lamberti, G. 2004. Restoration of an Indiana, USA, stream: bridging the gap between basic and applied lotic ecology. Journal of the North American Benthological Society 23:647–660.

Mulhouse, J., and Galatowitsch, S. 2003. Revegetation of prairie pothole wetlands in the mid-continental US: twelve years post-reflooding. Plant Ecology 169:143–159.

New, J. F., and Associates. 2003. Lake Station Mitigation Bank 2002 Monitoring Report. Walkerton, IN.

Nickum, M., Mazik, P., Nickum, J., and MacKinlay, D. 2005. Propagated fish in resource management. Fisheries 30:30–33.

Odum, H. T. 1989. Ecological engineering and self-organization. In Ecological Engineering: An Introduction to Ecotechnology, ed. Mitsch, W. J., and Jørgensen, S. E., pp. 79–101. New York: Wiley.

Qin, B., Xu, P., Wu, Q., Luo, L., and Zhang, Y. 2007. Environmental issues of Lake Taihu, China. Hydrobiologia 581:3–14.

Richards, R., Goldman, C., Frantz, T., and Wickwire, R. 1975. Where have all the Daphnia gone: the decline of a major cladoceran in Lake Tahoe, California-Nevada. Verhandlungen International Verein Limnologie 19:835–842.

Spieles, D. J. 2005. The role of biomanipulation in aquatic ecosystem restoration. In Progress in Aquatic Ecosystem Research, ed. Burk, A. R., pp. 59–82. New York: Nova Science Publishers.

Vallentyne, J. 1974. The Algal Bowl: Lakes and Man. Ottawa: Department of the Environment: Fisheries and Marine Service.

Vander Zanden, M., Chandra, S., Allen, B., Reuter, J., and Goldman, C. 2003. Historical food web structure and restoration of native aquatic communities in the Lake Tahoe (California–Nevada) Basin. Ecosystems 6:274–288.

Vannote, R., Minshall, G., Cummins, K., Sedell, J., and Cushing, C. 1980. The river continuum concept. Canadian Journal of Fisheries and Aquatic Sciences 37:130–137.

Warren, D., Bernhardt, E., Hall, R., and Likens, G. 2007. Forest age, wood and nutrient dynamics in headwater streams of the Hubbard Brook Experimental Forest, NH. Earth Surface Processes and Landforms 32:1154–1163.

Wetzel, R. G. 2001. Limnology: Lake and River Ecosystems. San Diego: Academic.

Chapter 10
Saltwater Ecosystems

Add some water to a terrestrial ecosystem and you can expect a boost in productivity. It is, after all, essential for life; most land-based creatures live their lives with only a small tolerance for desiccation. Now add more water, so that the ecosystem is periodically or permanently flooded. This is too much water for many organisms, and it will quickly bring them to the end of their capacity to live without oxygen. It may also restrict growth and development by accentuating the constraints of light attenuation, gas exchange, and nutrient assimilation. Now add salt, and you've complicated things with another level of stress. It's not that the marine and estuarine stress regimes preclude life, of course. On the contrary, some salt water ecosystems are disproportionately rich in biological diversity. It's just that salinity changes the rules of survival – and life in this environment requires first and foremost an adaptation to abiotic stress.

As we have seen, humans are quite adept at altering the stress regimes of the world around them. Much ecological stress can be attributed to human development and progress, but a curiously large amount is the result of human efforts to maintain ecosystems in a desirable state. Suppression of disturbance, for example, has historically been employed as a tool of ecosystem management though it ultimately leads to the decline of dominant species, catastrophic change, or the need for more intensive management. There is, too, a stress related to the management of ecosystems for the services they provide. This is most obvious with provisional services, not only through repeated harvest but also in the stocking and management of desirable organisms. Even regulating and cultural services can be ecologically stressful if they require an interruption or stagnation of succession.

So ecosystems are a function of their stress regimes, and human action – even well-intentioned – can add additional stress. Humans can also alleviate ecological stress, though it does not necessarily follow that the ecosystem will return to its pre-stress condition. Here, then, is a management choice: work to achieve the desirable ecosystem state in spite of the stress regime, or work to reduce anthropogenic stress and allow the ecosystem to respond accordingly. This is the stark dichotomy of holistic and individualistic management; the reality of most management efforts is probably somewhere in between these extremes. We have seen examples of such management in forest, grassland, and freshwater ecosystems. Now, with particular focus on natural and anthropogenic stress, we turn to the sea.

D. J. Spieles, *Protected Land*, Springer Series on Environmental Management, DOI 10.1007/978-1-4419-6813-5_10, © Springer Science+Business Media, LLC 2010

Intertidal Ecosystems

Life is particularly stressful between the tides. Alternating inundation and desiccation, strong currents, extreme salinity, and random buffeting by all manner of flotsam make for a harsh existence. As a result, the vascular plant community of the intertidal zone is comparatively low in diversity (Odum 1988). Salt marshes of the eastern and Gulf coasts of the United States, for instance, are dominated by just a few plant species, notably smooth cordgrass (*Spartina alterniflora*), saltmeadow cordgrass (*Spartina patens*) and black needlerush (*Juncus roemarianus*) in sharply zoned and nearly monotypic stands. On the other hand, the intertidal salt marsh is a region of great diversity for algae, invertebrates, and temporary fish and bird populations. The saltwater tide is the great force of the salt marsh. It can prohibit life, as in the barren pans of evaporating salt water. It can also enhance productivity as it brings nutrient- and oxygen-laden water in its advance and takes wastes and organic detritus in its retreat. There are other disturbances here as well: periodic coastal storm surge, displacement of organisms and resources by wave action, and even freshwater and nutrient influx from periodic flooding. Consequently, there is spatial and temporal patchiness within the strict tidal zonation (Sousa 1985). Where there is disturbance, there is succession, but the stress makes succession simpler in the salt marsh. There are few colonizers and – at least for vascular plants – a swift progression to a limited number of dominant communities. But it would be a mistake to characterize salt marshes as ecosystems of universal similarity, particularly if this assumption is based on only on dominant plants. Abiotic conditions and biotic responses vary within and among sites, and in this respect salt marshes share another commonality with forests, prairies, rivers and lakes: no two are alike (Adam 1990).

Community composition and action in the salt marsh are products of the stress and disturbance regimes. More often than not, human activity has added additional stress that has effected changes on ecosystem structure and function. Anthropogenic stress occurs in two broad categories: collateral damage and service exploitation. First, collateral damage. A great number of people live near the coast, and people in developed nations tend to have some problems with things like storm surges, mosquitoes, and even intertidal zones themselves. Consequently Americans have been impounding, ditching and filling salt marshes and constructing sea walls for centuries – historically to reclaim land for agriculture, and more recently for urbanization and development (Teal and Teal 1969; Gedan et al. 2009). Nearly as damaging have been the indirect effects of eutrophication, erosion, heavy metal contamination, freshwater diversion, and pesticide use.

Despite their inconvenient location and bothersome characteristics, salt marshes have some redeeming qualities. For one, they export large quantities of nutrients and organic matter to off-shore ecosystems. They also stabilize shorelines, buffer wave and storm surge action, and transform dissolved gasses and nutrients (Costanza et al. 1998; Beck et al. 2001). They are important breeding grounds and nurseries for a great diversity of fish, invertebrates, and birds, some of which are commercially valuable. This brings us to service exploitation. Provisional salt marsh services have been over-used, as in the harvest of commodities like fish, shellfish, clams and turtles. Other services have been systematically eliminated by humans or replaced with engineered structures (like seawalls). All of these alterations have effects on the salt marsh ecosystem – some expected, some unanticipated. If you add in the stresses associated with climate change and sea level rise, you have the makings for a new set of salt marsh regimes.

Salt Marsh Dieback and Restoration

Each of these stressors, natural and anthropogenic, alter the habitat suitability for and functional capacity of individual species. Recently, the accumulated stress on some salt marsh ecosystems appears to have reached a level of synergism sufficient to alter the dominant plant community. The change has become increasingly hard to dismiss. On the southeastern and Gulf coasts of the US alone, some 625,000 acres of salt marsh shows some degree of the phenomenon known as marsh dieback or die-off (Gedan et al. 2009). Dieback at some level of severity has reached most of the eastern seaboard in the last few decades. Characterized by a "browning" of marsh vegetation followed by nearly complete mortality of vascular plants, the progression can take several years or occur in a matter of months (McKee et al. 2004). The reasons for this are not entirely clear. The proximate causes are thought to be related to unchecked herbivory by snails, crabs, geese or nutria, perhaps accompanied by fungal infection. Ultimate causes that heighten consumer activity and hamper vegetation response may include climate-associated water logging or drought, variation in soil salinity, overharvest of predators, and even the overuse of fertilizers on distant terrestrial environments. It is a long list of potential triggers, and there isn't a simple series of events that is common to each case. Whatever the collection of causes, the result is a dramatic ecological regime shift. The decline of sediment-trapping plants like *Spartina* can lead to marsh erosion and subsidence, exposed mudflats in the intertidal zone, and encroachment of invasive species (Gedan et al. 2009). This new regime lacks some of the salt marsh services on which we have come to depend, and these services are not easily replaced.

Restoration of degraded salt marshes is a fickle business. The objective is typically the same as that of the many ecosystems considered in this book: restoration of historic conditions or the establishment of a particular desirable state. But the stress and disturbance that historically gave rise to the reference salt marsh state – or that allow for specific ecosystem services – are multifaceted and difficult to control. In the case of salt marsh dieback a convoluted set of local, regional and even global stressors present formidable obstacles to a designed restoration. Depending on the severity of the dieback event, the entire physical structure of the marsh may need to be reconstructed before biotic restoration can proceed.

Just such a project is underway in coastal Louisiana (Schrift et al. 2008), where a major dieback resulted in marsh erosion and subsidence. To allow for any sort of restoration, several areas of the marsh were impounded and amended with sediment dredged from the nearby bayou to build up the elevation and encourage revegetation. The technique has been reasonably successful: some treatment areas have revegetated rapidly, particularly those with remnant living vegetation that survived the dieback. So, salt marsh reconstruction may hold some promise for post-dieback recovery.

But can restored salt marsh be engineered to resemble the pre-dieback marsh? Perhaps, but there are three major difficulties. First, environmental conditions of the salt marsh have narrow parameters. In the study, less than 10 cm difference in elevation made the difference between rapid plant colonization and poor colonization. Clearly, elevation and hydrologic regime specifications are critical to the restoration plan. And, while elevation and hydrology can certainly be engineered, they will continue to change over time. Will continual engineering be required to maintain the desired characteristics? Second, the restored plant communities may not necessarily be the same as the pre-impact or reference communities. Results from the Louisiana reclamation show that the most successful sediment-amended marsh treatments had greater plant species richness than the reference condition, as might be expected in early succession. There are, therefore, a great many possible competitive outcomes. With precision management, the degree of convergence with reference conditions over time may be high, but in my opinion it is unnecessary and perhaps contrary to succession. To what management lengths are we willing to go to maintain the "typical" salt marsh? Finally, site-specific reconstruction is still subject to the stress and disturbance regimes of the greater region, including the factors that lead to degradation in the first place. Which of the critical proximate and ultimate causes of dieback can be minimized to protect the restored ecosystem, and which are beyond our control?

The dredge-and-fill reconstruction approach in Louisiana can be contrasted with salt marsh restoration in New England. These are areas that have not experienced dieback as such, but various flood control structures have long removed the marsh from the tide, and invasive species have become dominant. Restoration typically involves the re-establishment of a natural hydrologic regime. The idea is to allow the salt water to return to areas from which it had been excluded, thereby re-establishing a stress and disturbance regime that will select against brackish water-tolerant invasive species and encourage salt-tolerant species. This is accomplished primarily by the removal of tide impediments, the excavation of impoundments to restore flow, and the creation of pools in former ditches.

In a survey of 36 such restoration projects, Konisky et al. (2006) suggest that physical salt marsh features, like salinity and hydroperiod, can be restored to a reasonable approximation of the reference condition with relative ease, while the biological community does not comply so readily. Three years after restoration, the plant communities of restored salt marshes had not yet achieved equivalence with reference sites, while the animals that were used as restoration indicators were "indistinguishable among reference, impacted, and restored areas" (Konisky et al. 2006). Now, it may be that some ecosystem services were restored with physical modification alone, and that other services will follow with time. It may also be that the biotic restoration will eventually match reference conditions. But the dominant regimes have changed in subtle ways, and there is no reason to expect a complete return to the pre-impact state. The events leading to degradation, and also the alterations intended to promote recovery, are now contingencies of community assembly. And no two sets of contingencies are alike.

Subtidal Ecosystems

Move offshore from the intertidal zone and you'll find stress and disturbance regimes that are quite different. The same concepts of ecosystem structure and function apply, though it takes a bit more visualization on the part of the average land-lubber. There is a three dimensional structure to the subtidal ocean; it consists of sessile plants and animals, their substrate, and the associated vertical and lateral distribution of producers, consumers and decomposers. Boundaries are stress dependant and can be sharp or gradual, just as they can in terrestrial ecosystems. In fact, many of the same boundaries apply: light and oxygen availability, temperature, nutrient concentration, substrate quality. Subtidal communities are likewise patchy in time and space, with much heterogeneity derived from biogenic structures. These ecosystems are subject to the usual list of anthropogenic stressors, featuring eutrophication and overharvest, and are similarly facing a rather insidious future of climate change and acidification. Finally, there are disturbances that differ in scale and intensity. Hurricanes, wave action, thermal stress, atmospheric exposure, sedimentation and nutrient flux create patches that are available for colonization. The ensuing succession is entirely dependent on the colonizers and the conditions in which they find themselves (Connell and Keough 1985).

Certain ecosystems of the shallow ocean have attracted attention for their recreational appeal, extraordinary ecosystem services, and rate of change. Coral reefs might be the best example. Occupying only about 3% of the world's tropical continental shelf area, reefs are disproportionately rich in biodiversity and linked with a multitude of services, including direct fish harvest, storm abatement, sightseeing, and the provision of nursery habitat. Like salt marshes, reefs are sensitive to their physical environment. Rather specific tolerance limits for light, temperature, depth, salinity, and wave energy restrict the possible reef locations (Kaiser 2005). Human encroachment is adding another layer of restriction. Anthropogenic change in coral reef ecosystems can be abrupt and transformative, making it a prime example of alternate state theory (Folke et al. 2004).

The typical story of reef degradation goes something like this. Reefs in the state of coral dominance support a diverse community, the keystone of which is the colonial coral polyp and its symbiotic coralline algae. A product of the mutualistic relationship is calcium carbonate, which is deposited in vast quantities to form the structure of the reef. Coral reefs can grow for centuries but they are not permanent structures; sea level changes over geological time have driven many iterations of coral growth and destruction. Nevertheless, by the standards of a human life coral reefs can be ancient. They are also fragile, given the narrow conditions necessary for their formation and growth. They are, for instance, vulnerable to algal growth that can cover the coral and out-compete the coralline algae (Elmqvist et al. 2003). Excessive algae can be kept in check by herbivorous fish and invertebrates, but overfishing has in some cases reduced or eliminated algal herbivores. Released from grazing pressure, algae can reduce light penetration and eventually kill the coral. Aggressive algal growth can even prevent the colonization of coral larvae. As the algal growth outpaces herbivory, algae can replace coral as the dominant feature of the biotic community. In some cases, this scenario is accentuated by inland erosion and subsequent turbidity, which can kill coral, or by coastal eutrophication, which can encourage algal growth. Slight increases in water temperature and acidification, both associated with climate change, also place coral at a selective disadvantage and encourage the regime shift (Doney 2006). A condition known as coral bleaching – the white color signifying the demise of the symbiotic algae – is an indicator of stress and generally precedes a regime shift.

Humans, the harbingers of regime shift, are also working hard to reverse the process. Maunalua Bay in Hawaii is the site of a great regime battleground. Situated between Diamond Head and Koko Head, the picturesque bay and its watershed support a great number of permanent and temporary human residents. Not the least among its attractive features is the fringing reef that occupies much of the bay to about 1 km from the shore. The reef is in shallow water – only submerged about 1.5 m at highest tide – and it is substantially influenced by runoff from Oahu's Koolau Mountains. The runoff has long been a regulator of the reef's stress regime. The fresh water influences the salinity of the bay, and the sediment and nutrients carried by the runoff can affect light penetration. The reef has grown under these regimes for centuries, but until recently the terrestrial land cover, infiltration, and wetlands have minimized the sediment and nutrient load into the Bay.

Since the 1960s the stress in the Bay has changed. The watershed has rapidly urbanized, such that it is now about 45% developed. Much of the undeveloped land is steeply sloped and erosion-prone. The development has dramatically increased the impervious area of the watershed, the number of channelized and lined streambeds, and hence the amount of stormwater – along with sediment and fertilizers – routed directly into the Bay. The morphology of the Bay, particularly with the advent of seawalls and artificial channels, does not readily permit effluent to the open ocean; thus the sediment and its associated contaminants are trapped and continually re-suspended with wave and tidal action. Add in the occasional dose of heavy metal contamination and organic pollution, and you've got a recipe for an impaired body of water (Wolanski et al. 2009).

The impairment is so severe that the coral of Maunalua Bay is dying. Smothered by sediment, cut off from light, subject to severe spikes of freshwater, and over-enriched with nutrients, coralline algae productivity has declined precipitously. If ever there was a prime location for algal invasion, this is it. And – as though to ensure the shift to algal dominance – humans have long been overharvesting herbivores that inhabit the Bay, such as parrotfishes, surgeonfishes, manini and sea urchins. Sure enough, the invader has arrived, riding the wave of regime change. It have come in the form of mud weed (actually a colonial green algae), which was introduced to Hawaii 1981 and has since come to dominate Maunalua Bay. Within a few years it covered about 54 acres of former living reef and seagrass beds (Brostoff 1989). The invader earned its name by its ability to trap sediments, worms and mollusks in its blades, thus exuding a muddy excrement. Its dominance is so complete that conservationists have resorted to hand-removal in an attempt to save the reef. Over the last 3 years, volunteers have removed some 25 t of mudweed from Maunalua Bay.

Of course, mechanical removal alone will not be effective (McClanahan et al. 2000). The factors that shifted the stress regime will need to be addressed if algal dominance is to be broken. This is no small task. The entire inland and Bay infrastructure will need to be re-evaluated and altered, as will human behavior regarding fertilization, land use, and marine harvest. If all this can be accomplished, some new ecosystem may take shape in Maunalua Bay, perhaps one that does not feature mudweed. But there is no reason to expect that the Bay, having undergone such dramatic regime alteration, will ever return to its former state.

There is also nothing to suggest that there is only one inevitable alternative to the coral reef state. Dominance by invasive algae has received the most attention, but reefs under stress may also come to be dominated by sea urchins, sponges, sea anemones, anthozoans, soft (non-reef building) corals, or sea squirts. It is becoming apparent that "phase shift dynamics on coral reefs seem to be characterized by multiple drivers and multiple outcomes" (Norström et al. 2008). Myriad factors, including the nature of coral decline and the circumstances of invasion, are involved in coral reef regime shift. It makes no more sense to speak of the alternate stable state as a typical unit than it does to describe coral reefs as all being basically alike.

Artificial Reefs

Another approach to habitat restoration has been attempted at Maunalua Bay, as it has in other shallow marine environments around the world. Like salt marsh construction with dredge material or the re-meandering of rivers, it is from the "Field of Dreams" school of thought: if you build it, they will come (Palmer et al. 1997). The Maunalua Bay Artificial Reef Site was created in 1961 to enhance the existing reef habitat, primarily by increasing the productivity and diversity of fish in the Bay. Despite the shady history of ocean refuse disposal in the name of habitat enhancement (Buckley 1982), the Maunalua Bay project appears to have been accepted for nearly 40 years as an ecological improvement. Initially the artificial reef consisted of some 1,600 stripped automobiles that were dumped into the bay about 2 km from shore, south of the natural reef. This was followed by 2,100 t of concrete pipe, a few scuttled barges, several hundred automobile tires, and numerous concrete modules (Brock and Norris 1989). Concrete block deployment continues today. These artificial ecosystems are not intended to replace the natural reefs; rather they are constructed to add habitat heterogeneity and augment the biotic community of the Bay. But artificial structures have been used in reef restoration elsewhere. Could this approach be used in the restoration of the Maunalua Bay reef? There appears to be some potential for the use of artificial substrates in the attraction of coral larvae and the establishment of fish assemblages, though such work is in its infancy. But it is clear that reef restoration cannot be accomplished with engineering alone: "reduction of pollution, limits to fishing, regulation of coastal development, and dealing with both human population growth and consumption of natural resources" are prerequisite (Seaman 2007). And even if they are accompanied by stress reduction, artificial reefs are the epitome of the designer ecosystem – they are a static surrogate for a dynamic system.

Deeper Water

The stress regime of the open sea is less influenced by terrestrial processes than that of salt marshes or estuarine bays, but it is still the governor of biotic distribution and abundance. Deep-water pelagic ecosystems are not particularly amenable to direct observation, but we know enough to say that these are spatially heterogeneous and temporally transitional systems with complex trophic structure (Kaiser 2005). The food web may at times be regulated from the top by predator activity, but at other times it is bottom-up nutrient availability that drives production. Variability in time and space is largely related to fluctuations in stressors such as light, salinity, temperature, density, nutrient and oxygen availability, and to corresponding biotic responses. Disturbances include upwellings, eddies, gyres, and horizontal currents, front formation and mobility, as well as periodic changes in any components of the stress regime. The resulting ecosystem is multifaceted, highly changeable, and vast. Given their size, complexity, and distance from shore, it would seem that ecosystems of the open ocean may be somewhat insulated from anthropogenic stress.

They are not. Overfishing is the most direct human encroachment on pelagic ecosystems, and it is continuing on a staggering scale. Eighty to 100 million tons of pelagic fish are harvested worldwide each year, with perhaps a quarter of the total discarded as by-catch and another quarter harvested illegally (Pauly et al. 2003). Every indication suggests that this rate of harvest is unsustainable. Indeed, the annual world catch has plateaued and declined slightly since 1980, forcing the harvest into deeper waters and lower on the food chain. By 2000, for example, approximately 40% of the world's marine fish harvest occurred at a depth of 1,000 m – nearly double the proportion caught at this depth in 1950 (Pauly et al. 2003). There is no mystery here. The depths are increasingly being harvested because surface waters are in a state of near-complete exploitation. This is an ecological stress of monumental proportions, and coupled with the ramifications of climate change it constitutes a regime shift.

We have experienced regime shifts in the deep ocean before. In the north Pacific, for example, the pelagic community underwent a dramatic shift in 1977 and again in 1989 (Hare and Mantua 2000). These shifts appear to have been climatically driven at least in part, but it is becoming more apparent that harvest-driven changes in competition and predation can play a role (Scheffer et al. 2001). Whatever their cause, the shifts have profound implications for fisheries management. In the 1977 shift, Alaskan salmon populations increased, but there were declines in Alaskan shrimp, salmon, and oyster populations. According to Hare and Mantua (2000), the 1989 shift was characterized by declines in "western Alaska chinook, chum, and pink salmon, British Columbia Coho, pink and sockeye salmon, west coast salmon, and groundfish recruitment" along with a surge in Bering Sea jellyfish. Both shifts appear to have been a response to changing disturbance and stress regimes; as anthropogenic pressures and abiotic conditions continue to change it is likely that community shifts will continue. And there is no reason to expect the shifts to be simple oscillations between two possible states; as we have seen in other ecosystems, there are multiple community responses to the array of abiotic possibilities in the ocean environment.

Thus it seems that there is no single state for pelagic ecosystems (or any other, for that matter). There are no normal and appropriate lists of species and relative abundance, no typical set of environmental conditions. To be sure, there are finite possibilities of the types of species that might exist in certain conditions, but the variable permutations of climate, current, and the host of other biotic and abiotic factors provide for a wide variety of ecosystem conformations. Furthermore, the *rate* of change in these multiple parameters is itself variable; thus some shifts in stress regime may be gradual while others are abrupt. Given this dynamic range of stress and response, it seems unlikely that a particular patch of ocean is limited to only a few "domains of attraction." Rather, the alternate states or regimes that we identify are nothing more than ephemeral patterns in a sea of environmental variability.

At least this is the individualistic view of regime shifts. Others have conceptualized regime shifts and alternate states more holistically. Depending on one's perspective, the management of pelagic ecosystems might feature different goals and actions. For instance, if regime shifts are seen as fluctuations between a limited number of alternate states – one of which is the desirable state – management would seem to be a matter of determining and maintaining a stress regime suitable for the desired dominant species. This accomplished, the rest of the ecosystem, within some range of variability, should follow suit. On the other hand, an individualistic perspective might yield a management scheme that has no stable state in mind. The inherent variability of abiotic conditions and biotic responses would suggest a multitude of alternate states, rendering the categorical concept of different states somewhat meaningless. This is not to suggest that the individualist would not recognize desirable conditions, for presumably even individualists eat and enjoy clean water. The achievement of desirable conditions, though, would not be a matter of maintaining a particular community. Instead, it might focus on limiting or eliminating anthropogenic stress and on rejecting the perception that desirable functions may only be associated with a single ecological state.

Which model have we followed in the management of our pelagic ecosystems? Well, first of all, the term "management" must be used somewhat loosely here. In comparison with terrestrial and even freshwater ecosystems, we can hardly control any aspects of the deep marine environment. In some respects we may be changing the open oceans, as in the case of climate change, acidification, and ice melt-driven changes in currents – but though they are human caused, these too are woefully beyond our ability to manage on the short term. The one thing we can manage is harvest (Kaiser 2005). Long unmanaged, many marine fisheries came under national jurisdiction and international agreement in the mid- to late-twentieth century. This has led to assessment and management of commercially viable species based on models of recruitment, growth, harvest, and mortality. Throughout most of the history of fishery management the goal has been the optimal yield of single species stocks, primarily with the use of catch limits. This method of management is entirely based upon the notion of the stable ecosystem. All other things being equal, the reasoning goes, we should be able to harvest this many fish at this size range and still allow a viable population for future harvest. But all other things are never equal – stress and disturbance regimes and their biotic responses have a way of changing in unpredictable ways. Meanwhile, social and political pressures continue to push the limits of fishery production. As a result, quota-based "maximum sustainable yield" management has not prevented the demise of marine fish stocks.

With mounting evidence of immanent fishery collapse and general damage to marine communities, fisheries managers are beginning to embrace the ecosystem approach. This means different things to different people, but it can include such practices as the creation of reserves or no-catch areas, the reduction of by-catch, or the timing of harvest to minimize the impact on ancillary species. Conceptually, the ecosystem approach to fisheries management is an interesting hybridization of strong and weak holism; it is truly the "unit" view of the ecosystem struggling to move along the continuum toward individualism. Let me illustrate this with a recent article called the "Ten Commandments for Ecosystem-Based Fisheries Scientists" by Francis et al. (2007). As "an effort to accelerate the ongoing paradigm shift in fisheries science from the traditional single-species mindset toward more ecosystem-based approaches" one would expect the commandments to reflect current concepts of the ecosystem in this field. The first commandment – keep a perspective that is holistic, risk averse, and adaptive – relates a view of holism that is decidedly weak. It takes into account "the constantly changing climate-driven physical and biological interactions in the ecosystem" including dynamic trophic interactions and adaptability. Likewise, the 8th commandment – account for ecosystem change through time – hints at the likelihood that these ecosystems have not always been and will not forever be as they are now. Yet commandments 3–7 suggest that the marine ecosystem is a unit that can be preserved in the ideal state if we only heed a few basic details. They are, in order: maintain old-growth age structure in fish populations; characterize and maintain the natural spatial structure of fish stocks; characterize and maintain viable fish habitats; characterize and maintain ecosystem resilience; and identify and maintain critical food web connections. To be fair, many of these are aimed at curtailing careless practices that threaten fish stock viability, and they are – in the weak sense of holism – concerned with factors beyond species of commercial interest. These are worthy proposals. But they also include a healthy dose of holism in the strong sense, intimating that the "stability domain of the existing food web" can be perpetuated indefinitely (Francis et al. 2007).

I know why the ecosystem-as-self-perpetuating-unit notion persists in fisheries management, and it is not because managers think of the ecosystem as a superorganism. It is the same reason that foresters seek stability on tree farms and city managers regulate river flooding. In a word, the reason is *service*. This is the classic dilemma of ecological conservation: how do we maintain constancy of yield, function, and service in systems that are not constant in space or time? It certainly can be done – a field of corn is an example of a highly ordered ecosystem with consistent service. Fisheries management, even in the ecosystem approach, follows the same path: what aspects of the ecosystem can we manipulate to ensure a reliable harvest or predictable function?

Humans have been managing ecosystems to this end for thousands of years, and I understand the desire for constancy (though I have no illusions that such systems are indefinitely sustainable). I do question, however, the extension of strong holism to ecosystems under protection as natural areas for their regulating, supporting, or cultural services. In this chapter we have seen instances of management for the ideal design of salt marshes and reefs; they bear resemblance to efforts to achieve and maintain the ideal freshwater wetlands, rivers, lakes, grasslands and forests. In many of these examples we see great adherence to commandments 3–7 as described above, with an overarching goal of achieving and preserving the ecosystem in the state we desire. But the 8th commandment, I believe, is greater than the rest, for ultimately the individual organisms of an ecosystem will respond differentially to changes in disturbance and stress regimes. At that point, the unit of the ecosystem as we perceive it will cease to be – it never really existed anyway – and a new organization will temporarily emerge. This is happening now and will continue repeatedly, quite independent of human preferences.

References

Adam, P. 1990. Saltmarsh Ecology. Cambridge: Cambridge University Press.

Beck, M., Heck, K., Able, K., Childers, D., Eggleston D., Gillanders, B., Halpern, B, Hays, C., Hoshino, K., and Minello, T. 2001. The identification, conservation, and management of estuarine and marine nurseries for fish and invertebrates. BioScience 51:633–641.

Brock, R., and Norris, J. 1989. An analysis of the efficacy of four artificial reef designs in tropical waters. Bulletin of Marine Science 44:934–941.

Brostoff, W. 1989. Avrainvillea amadelpha (Codiales, Chlorophyta) from Oahu, Hawaii. Pacific Science 43:166–169.

Buckley, R. 1982. Marine habitat enhancement and urban recreational fishing in Washington. Marine Fisheries Review 44:28–37.

Connell, J. and Keough, M. J. 1985. Disturbance and patch dynamics of subtidal marine animals on hard substrata. In The Ecology of Natural Disturbance and Patch Dynamics, ed. Pickett, S., and White, P. S., pp. 125–152. Orlando: Academic.

Costanza, R., d'Arge, R., De Groot, R., Farber, S., Grasso, M., Hannon, B., Limburg, K., Naeem, S., O'Neill, R., and Paruelo, J. 1998. The value of the world's ecosystem services and natural capital. Ecological Economics 25:3–15.

Doney, S. 2006. The dangers of ocean acidification. Scientific American 294:58–65.

Elmqvist, T., Folke, C., Nyström, M., Peterson, G., Bengtsson, J., Walker, B., and Norberg, J. 2003. Response diversity, ecosystem change, and resilience. Frontiers in Ecology and the Environment 1:488–494.

Folke, C., Carpenter, S., Walker, B., Scheffer, M., Elmqvist, T., Gunderson, L., and Holling, C. 2004. Regime shifts, resilience, and biodiversity in ecosystem management. Annual Review of Ecology, Evolution, and Systematics 35:557–581.

Francis, R., Hixon, M., Clarke, M., Murawski, S., and Ralston, S. 2007. Ten commandments for ecosystem-based fisheries scientists. Fisheries 32:217–233.

Gedan, K., Silliman, B., and Bertness, M. 2009. Centuries of human-driven change in salt marsh ecosystems. Annual Review of Marine Science 1:117–141.

Hare, S., and Mantua, N. 2000. Empirical evidence for North Pacific regime shifts in 1977 and 1989. Progress in Oceanography 47:103–145.

Kaiser, M. J. 2005. Marine Ecology: Processes, Systems, and Impacts. New York: Oxford University Press.

Konisky, R., Burdick, D., Dionne, M., and Neckles, H. 2006. A regional assessment of salt marsh restoration and monitoring in the Gulf of Maine. Restoration Ecology 14:516–525.

McClanahan, T., Bergman, K., Huitric, M., McField, M., Elfwing, T., Nyström, M., and Nordemar, I. 2000. Response of fishes to algae reduction on Glovers Reef, Belize. Marine Ecology Progress Series 206:273–282.

McKee, K., Mendelssohn, I., and Materne, M. 2004. Acute salt marsh dieback in the Mississippi River deltaic plain: a drought-induced phenomenon? Global Ecology and Biogeography 13:65–73.

Norström, A., Nyström, M., Lokrantz, J., and Folke, C. 2008. Alternative states on coral reefs: beyond coral-macroalgal phase shifts. Marine Ecology Progress Series 376:295–306.

Odum, W. 1988. Comparative ecology of tidal freshwater and salt marshes. Annual Review of Ecology and Systematics 19:147–176.

Palmer, M., Ambrose, R., and Poff, N. 1997. Ecological theory and community restoration ecology. Restoration Ecology 5:291–300.

Pauly, D., Alder, J., Bennett, E., Christensen, V., Tyedmers, P., and Watson, R. 2003. The future for fisheries. Science 302:1359–1361.

Scheffer, M., Carpenter, S., Foley, J., Folke, C., and Walker, B. 2001. Catastrophic shifts in ecosystems. Nature 413:591–596.

Schrift, A., Mendelssohn, I., and Materne, M. 2008. Salt marsh restoration with sediment-slurry amendments following a drought-induced large-scale disturbance. Wetlands 28:1071–1085.

Seaman, W. 2007. Artificial habitats and the restoration of degraded marine ecosystems and fisheries. Hydrobiologia 580:143–155.

Sousa, W. 1985. Disturbance and patch dynamics on rocky intertidal shores. In The Ecology of Natural Disturbance and Patch Dynamics, ed. Pickett, S., and White, P. S., pp. 101–124. Orlando: Academic.

Teal, J., and Teal, M. 1969. Life and Death of the Salt Marsh. New York: Ballantine.

Wolanski, E., Martinez, J., and Richmond, R. 2009. Quantifying the impact of watershed urbanization on a coral reef: Maunalua Bay, Hawaii. Estuarine, Coastal and Shelf Science 84:259–268.

Chapter 11
Protecting the Shifting Quilt

If there is one point on which the holist and the individualist can agree, it is that humans have exerted an unprecedented intensity and variety of stress on ecological systems. This is regrettable because it jeopardizes ecosystem services, threatens the quality of human life, and ultimately influences the number of humans our planet can support. It has also pushed numerous non-human species to and past the brink of extinction. Anthropogenic stress has intensified in the past century, even as we have struggled to understand these things we call ecosystems. Our recognition of ecological crisis – late in the game, perhaps, but in earnest – has generated questions of ecosystem status and performance. In the United States and around the world certain aquatic and terrestrial ecosystems have been placed under various levels of protection from human encroachment and development. But protection often seems inadequate. Protection will not immediately negate a history of degradation, nor will it necessarily reverse current anthropogenic stress. And so we feel compelled to act. What can we do to define and achieve recovery in degraded ecosystems?

Our course of action, ranging from a minimalist "leave nature to her devices" approach at one extreme to intensive management on the other, may well depend on how we conceptualize the system we are trying to protect. In this book, I have portrayed the range of ecosystem conceptualization as a continuum from the holistic, self-sustaining climax unit to the individualistic, radically contingent coincidence. Where on this continuum is the prevailing thought in the United States, and how has it informed our ecosystem management efforts?

In previous chapters I have described some major ecosystem management projects in a variety of environments. We have seen examples of forests that are managed with water level manipulation and fire to arrest succession or to re-direct it to a desirable state. Some grasslands, too are managed for stability in spite of succession, as fire, herbicides, grazing, mowing and mechanical removal are used to prevent the encroachment of woody species. Mitigation wetlands are constructed and restored according to a code of design, and deviations are considered unsuccessful ecosystems. We engineer the meanders back into streams as we once engineered them out, and then we expect the pre-impact biota to return in their characteristic numbers. When biological communities don't take the form we desire, we introduce species – native or not – to attain a productive lake, to create a fishable stream, to achieve the

D. J. Spieles, *Protected Land*, Springer Series on Environmental Management,
DOI 10.1007/978-1-4419-6813-5_11, © Springer Science+Business Media, LLC 2010

ideal ecosystem. Saltwater systems are similarly being built and managed according to specifications, and where we can't resist change by construction we attempt do so with regulation.

All of this is indicative of a national ecosystem mindset that is far to the holistic side of the continuum. We seek to maintain our ecosystems as units – to keep them in the desirable domain of attraction. We do this to protect particular ecosystem services and to preserve ecological legacies, but in many cases that provide neither commodity nor cultural significance we appear to be managing the ecosystem to the state we think it *should* be. The practices by which we direct ecosystems to and hold them in the desired state – controlling disturbance, colonization, and competition – are drawn directly from Clementsian ecology (Pickett et al. 2009). This mindset is a central tenet of American ecosystem management, and it is deeply engrained in our national policy. The interagency ecosystem approach, wetland mitigation rules, agency procedures, national and state stocking programs, federal stimulus expenditures, conservation legislation, and even the IRS definition of conservation are all aimed at maintenance of the ideal ecosystem unit. Many private and nonprofit conservation organizations have followed suit.

And yet in every case, from microbial ecosystems to million acre forests to deep ocean environments, we see ecosystem characteristics that can only be described as individualistic responses in loose and temporary association. We find that we can't objectively define the proper domain of attraction for any ecosystem because each ecosystem at any given time is a composite of patches in spatial and temporal flux. Experience has shown that restoration of abiotic characteristics is no guarantee that the ecosystem will revert to any domain of attraction. On the contrary, examples like the cessation of logging in the Otter Creek wilderness, post-eruption succession on Mount St. Helens, forest encroachment on Ebey's Landing, and the inability to replicate pre-impact communities in restored wetlands and streams all suggest that the archetypal self-sustaining unit is not readily achieved. This casts real doubt on the conceptual domain of attraction. In fact there is abundant evidence that ecosystem "domains" are not categorical states but temporary patterns in a world of continuous change.

It is true that organisms of similar adaptation may respond to environmental change in similar ways. It is also true that some biotic activity is reinforced by symbiosis and feedback mechanisms. From these points, though, it is quite a leap to conclude that biological communities gravitate toward discrete modes of structure and function. If ecosystems are stability-seeking units, why are they so dynamic? If they are all attracted to certain domains, why are they all so unique? If they are inherently resilient, why are they so difficult to restore and maintain in the ideal state?

The ecosystem-unit, domain-of-attraction mindset only makes sense if we are sufficiently vague in our characterizations and arbitrary in our definitions. The vagueness includes ecosystem recognition based on dominant species and three-dimensional structure. Thus we can compare a beech-maple forest with an oak-hickory forest, and we can evaluate what we perceive to be exemplary or degraded states of each. In reality, of course, no two stands of beech-maple forest are alike, and I have encountered portions of beech-maple forest that harbor neither beech nor maple trees.

But in our desire to categorize we tend to trivialize and minimize differences, to chalk them up as being within the range of variation for that type of ecosystem. If we can protect the proper conditions for the dominant species, the mindset dictates, the composition, structure and function of the rest of the ecosystem will fall into place. This way of thinking is a symptom of our historical roots; it is the legacy of the superorganism. All too often, we manage ecosystems as though they are ill and we want to return them to health. Unfortunately our prevailing concept of health is based on a unit-view of the ideal ecosystem that is, or was, a unique response to a unique regime. In many cases the ideal state can no longer be achieved, and if achieved it cannot be maintained. But the mindset is persistent, and it appears to be the main impetus for modern American ecosystem management.

Holism, Strong and Weak

Even as mounting evidence has suggested that ecosystems do not occur in units, human manipulation of stress and disturbance has continued to select for some rather undesirable ecological manifestations. Understanding as we do that desirable species and useful functions do not occur in isolation, we have often sought to restore, preserve and protect the characteristics that make desirable end products possible. In this way we attempt to manage the stream so it will keep producing trout, the savanna so it will support the rare butterfly, and the prairie so that it resembles its historic conditions. These actions are based on holism in the strong sense – that the ecosystem unit has a mature or optimal state toward which it should progress and to which it should return after disturbance. But in effect our well-intended preservation efforts are attempts to force a stable equilibrium in a non-equilibrium world. Each instance of historic and current anthropogenic stress and disturbance is indelibly incorporated into the ecosystems we seek to protect today. Furthermore, conditions of *natural* stress and disturbance are not constant – they change over ecological and geological time. Nor are conditions static in the ecosystem's landscape matrix. All of this means that a return to ecosystems that existed before European-American contact is an unachievable goal. In any case, the ecosystems at the time of contact had already been sculpted by thousands of years of Native American stress and disturbance regimes, and those ecosystems were not stable units either. The same applies to *any* idealistic ecological unit. It cannot be created and held at equilibrium in the midst of a dynamic landscape and fluctuating environmental conditions. The fact that biota themselves contribute to environmental change is further reason to reject the practice of arresting an ecosystem at some ideal state in the name of conservation.

So holistic management in the strong sense is problematic – does this mean that the concept of holistic management should be discarded altogether? Certainly not; it only means that our operational definition of holism needs to be adjusted to the "weak" sense of the term. Ecosystem management can and should consider multiple factors, species, and processes as being integrated but also transitional. Spatial and temporal boundaries would necessarily be blurred in the weak version of holism; since there is no discrete "unit" it would make little sense to think of organisms or functions being part of or external to the system. Such arbitrary boundaries are not nearly as useful as the real boundaries of physicochemical conditions that are conducive to ecological processes. Best of all, a weak view of holism corresponds nicely with our emerging understanding of how ecosystems work. Rather than repeated management to maintain an ecosystem state that just won't seem to behave as we think it should, we would instead be managing stress and disturbance regimes to encourage response and undirected succession.

Such a shift in management mindset would require us to relinquish some control and accept a certain level of unpredictability. Typically, ecosystem management is based on the fear of change: if we don't burn/mow/spray/introduce/remove/regulate, then we might lose this species or that function and we may experience an invasion of undesirable species. That some species and functions are imperiled is true enough, and as noted above, it is regrettable. I have no wish for the extirpation of the Karner blue or the wood stork. I also can readily acknowledge that certain ecological functions and critical services are impeded in a landscape of stress. I do not, therefore, oppose ecosystem management. What I oppose is management for some perceived ideal state, particularly if it is based on historical arrangement, when that conformation is contrary to the existing conditions of stress regime and constituent species. I suggest that we can manage ecosystems holistically (in the weak sense) solely by managing anthropogenic stress. This might mean reducing cultural eutrophication, overharvest, and fragmentation; it might also mean re-instituting disturbance regimes and stemming the tide of imported exotic species. But we can't undo stresses of the past, and we can't eliminate exotic species that have already been introduced. These are now part of each ecosystem's set of historic contingencies. The best action we can take now for the long term viability of a protected ecosystem is to reduce stress where possible and let the system mature, release, and reorganize – warts and all.

The Four Horsemen Revisited

In the first chapter I raised questions about prevalent aspects of American ecosystem management. These central tenets of ideal ecosystem characteristics (integrity and health) and resistance to change (stability and resilience) have arisen directly from the school of strong holism. In the strong holism, ecosystem-unit approach, integrity and health describe characteristics of the ideal ecosystem state, the desirable domain of attraction. Stability and resilience are concerned with the maintenance of that state by the ecosystem itself. But like holism, the concepts of integrity, health, stability and resilience can be defined and interpreted in more than one way. Each concept can be a worthy and achievable goal of ecosystem management if it is nudged down the continuum:

> Ecological Integrity (strong version) – preserving native species populations, in their characteristic numbers, with their evolved or historic interactions

(weak version) – the presence of spatiotemporal patchiness and functional diversity

Ecosystem Health (strong version) – presence of conditions favorable for the optimal functioning of the ecosystem

(weak version) – presence of the capacity for response, successional change, and migration at different scales

Stability (strong version) – the constancy of an equilibrium state in the face of an external perturbations

(weak version) – sufficient scale-appropriate time to allow for response between disturbance events or shifts in stress regime

Resilience (strong version) – the amount of change the system can undergo and still retain the same controls on function and structure; the degree to which the system is capable of self-organization

(weak version) – the capacity for post-disturbance reorganization

Resilience, stability, integrity, and health in their "strong" forms are describing an ideal ecosystem that does not exist. Or, more correctly, it is one that exists only in the eye of the beholder within carefully defined spatial and temporal boundaries. Ecosystem managers have demonstrated that ideal ecosystem characteristics can be forcibly maintained over years or even decades. But over longer time scales, or under conditions further removed from the desired regime, preservation or restoration of the ideal progressively becomes more labor intensive, more likely to require species introduction and removal, and more difficult to maintain. Ultimately, as such effort brings its own suite of ecological stressors, it can become counterproductive.

On the other hand, the weak interpretation of these terms would suggest a management scheme based not on three-dimensional structure or desirable dominants but instead on stress, disturbance, and response. This is not a call for a hands-off, zero management approach. Rather, it is a shift in management goals and activities. It is based on the recognition that ecological function and ecosystem services are inhibited by anthropogenic stress, not by successional change. Management solely for the reduction of anthropogenic stress would require us to abandon the designer approach and even the holistic implications of self-design. Instead, we would need to accept that ecological change happens according to no design at all.

Along the Continuum

The ecosystem management examples described in this book are not all neatly categorized as holistic or individualistic. In their breadth and variety we can see the schemes and outcomes along the continuum. Holism in the strongest sense is the management style of choice at our sanctuaries, including places like Corkscrew Swamp and Curtis Prairie. These are demonstration ecosystems, operated with curatorial precision. But strong holism is less excusable in a great many other projects: the Kissimmee River is representative of systems that are managed for the achievement and maintenance of one ideal state. As I have noted, this attitude is often driven by federal regulation (as in wetland mitigation), federal dollars (Maunalua Bay), and federal tax law (land trusts). Some management projects have effectively become holistic in the weak sense, much to the dismay of holistic preservationists or restorationists. The Boundary Waters Wilderness, Six Rivers National Forest, and our pelagic fisheries are examples of ecosystem management that has fallen back to a position of stress reduction after an inability to prevent change. And there are even some examples of management that is individualistic, or nearly so: Otter Creek is typical of the stress-reduction, open-succession model of many National Wilderness Areas; the Mount St. Helens National Volcanic Monument is being left, wilderness-style, to undesigned succession; restoration of tidal influence is the major management tool on the salt marshes of the Gulf of Maine; Hubbard Brook and its tributaries and watersheds are experimentally manipulated but have no design template. Notably, these last examples are able to be functional, attractive, and even beneficial ecosystems despite their lack of design.

Of course, American ecosystem management projects occur all along the continuum. In my estimation a growing number are adopting an ecological version of laissez faire; whether this is a new appreciation of Gleasonian ecology or simply due to budget cuts I cannot say. In any case I would argue that it is time for the "new" nonequilibrium ecology (which is no longer, if it ever was, new) to be incorporated into an approach to ecosystem management that embraces dynamic variability.

This is already happening. Let me cite just one example (Institute for Natural Resources 2009). The Oregon Department of Forestry in cooperation with the Institute for Natural Resources and the US Forest Service initiated a Dynamic Ecosystem Project in 2007. The project is aimed at reconciling natural resource policy and regulation with current management practices and "the scientific understanding of ecosystem dynamics." Appropriately called "Ecosystem Dynamics Management," the concept seeks to implement a number of points that I would place on the individualistic to weakly holistic part of the continuum:

- The non-equilibrium nature of ecosystems is not just an inconvenience that should be tolerated, it is an essential driver of system structure and productivity.
- Fixed rules for protecting certain values or producing commodities may lead to serious disruptions in the long term delivery of desired conditions, goods and services.
- A single landscape pattern will not persist except over large areas at multi-century time frames.
- Management should be designed to accommodate uncertainty by accomplishing multiple alternative goals.
- Management should focus on rehabilitation and maintenance of ecosystem functions rather than maintaining a particular species of physical appearance.
- Fundamentally, managing in an ecosystems dynamic framework involves development of management techniques tailored to the diverse ecosystem dynamics at work on different sites.

Dynamics Management is at the initial stages of planning and implementation in Oregon, but it has intriguing possibilities. Some examples: The establishment of "regime standards" at broad geographic and temporal scales, so that ecological stress factors like water quality can be assessed over a broad range and time, not at a point. Managing for heterogeneity instead of sharp boundaries among upland and riparian forests, acknowledging that the boundary between aquatic and terrestrial systems is blurry. Intentionally providing material (e.g. coarse woody debris) to prime the next disturbance event (e.g. movement of woody debris into the riparian corridor). Minimizing stream stress from roads. Creating a less dense and more heterogeneous forest with respect to age and species diversity by varying harvest and reforestation techniques. Encouraging a range, not a uniformity, of disturbance processes.

None of these ideas are intended to manage for the ideal, stable ecosystem. Instead, they are aimed at promoting long term function by incorporation of dynamic variability into the management plan. There are challenges, to be sure. People – including consumers, foresters, environmental advocates, scientists, and anyone else that has anything to do with the ecosystem – often expect stability and uniformity, not dynamic variability. It will take education, explanation, and demonstration to convince all stakeholders that change must be embraced. This will be difficult in the short term. In the long term, ecosystems will change, with or without our approval.

Protecting the Shifting Quilt

"If you believe that nature is a continually shifting quilt of patches, then there's no order, and why bother about conservation?" (Chaffin 1998). Such were Eugene Odum's thoughts on non-equilibrium ecology. The implication is that we should only be interested in protecting orderly things. Based on the current understanding of ecosystems and on the outcomes of recent ecosystem management, I respectfully reject the premise of Odum's question. Human perceptions of ecological order – the stable climax community, functional equilibrium, persistence amidst disturbance and regime shift, appropriate species in their characteristic numbers – are not prerequisite for conservation. On the contrary, our ecosystem protection efforts must allow for dynamic unpredictability, for that is the nature of the systems we seek to protect. It is also, I think, a fallacy to suggest that individualistic ecology is incompatible with the protection of critical habitat or ecosystem services. It is incompatible with *static* habitat and *fixed* ecosystem services. But this shift in our framework need not mean that we abandon the old growth or that we surrender our rare and endangered species to the weeds of the world. Individualistic ecosystems still have a place for late successional associations. They still have room for rare environments and diverse biota. They still have aggregate functions on which humans depend. They always have. Nor does acceptance of the dynamic approach mean that we must lose the ground that John Muir, Gifford Pinchot, Aldo Leopold, Eugene Odum and many others have gained in the struggle for conservation. What it means is that our protection and management effort should no longer be fixated on what an ecosystem was or is; rather it should be intended to allow an ecosystem to become what it may.

References

Chaffin, T. 1998. Whole-earth mentor. Natural History 107:8–10.
Institute for Natural Resources. 2009. Dynamic Ecosystem Project Summary. Corvallis: Oregon State University.
Pickett, S., Cadenasso, M., and Meiners, S. 2009. Ever since Clements: from succession to vegetation dynamics and understanding to intervention. Applied Vegetation Science 12:9–21.

Index

CPSIA information can be obtained at www.ICGtesting.com
233786LV00011B/38/P

9 781441 968128